SOIL TO FOIL

SOIL TO FOIL

ALUMINUM AND THE QUEST FOR INDUSTRIAL SUSTAINABILITY

SALEEM H. ALI

Columbia University Press   *New York*

Columbia University Press
Publishers Since 1893
New York   Chichester, West Sussex
cup.columbia.edu

Library of Congress Cataloging-in-Publication Data
Names: Ali, Saleem H. (Saleem Hassan), 1973– author.
Title: Soil to foil : aluminum and the quest for
industrial sustainability / Saleem Ali.
Description: New York : Columbia University Press, [2023] |
Includes bibliographical references and index.
Identifiers: LCCN 2022050229 (print) | LCCN 2022050230 (ebook) |
ISBN 9780231204484 (hardback) | ISBN 9780231555562 (ebook)
Subjects: LCSH: Aluminum. | Aluminum industry and trade. |
Mineral resources conservation.
Classification: LCC TA480.A6 A339 2023 (print) | LCC TA480.A6
(ebook) | DDC 669/.722—dc23/eng/20221206
LC record available at https://lccn.loc.gov/2022050229
LC ebook record available at https://lccn.loc.gov/2022050230

Cover design: Julia Kushnirsky

To Larry Susskind, mentor for over two decades, who taught me to always keep my eye on seeking solutions that link science with society through a process of patient consensus building

As in digging for precious metals in the mines, much earthy rub-bish has first to be troublesomely handled and thrown out; so, in digging in one's soul for the fine gold of genius, much dullness and common-place is first brought to light.

—Herman Melville, *Pierre; or, The Ambiguities*

CONTENTS

PREFACE

Human progress is often defined by how well particular elements of the earth are harnessed: the Bronze Age, the Iron Age, and the Silicon Age are common delineations of history. Yet the most abundant metal in the earth's crust, which has played a pivotal role in myriad technologies and inventions from aircraft to soda cans—aluminum—has largely been neglected in such lofty conversations about technological development. *Soil to Foil* aims to fill this gap in the popular earth science/environmental studies literature but also seeks to use the parable of aluminum for broader lessons on sustainable production and consumption of "nonrenewable resources." The book explores how the scarcity and abundance of natural resources are contested concepts based on energy investments in extraction, the modularity/durability of products, and the eventual circularity of material flows. The story of aluminum's presence in human technologies is thus a corollary for a broader conversation about how science and industry have challenged natural limits and how we can more sustainably manage the elemental resources of our planet.

Inspiration for this manuscript comes from the works of Mark Kurlansky, whose books on specific resources such as *Cod*, *Salt*,

Salmon, Oysters, Paper, and *Milk* have provided a marvelous focal point for broader historical narratives about the trials and triumphs of human development. While Kurlansky's lens is largely historical, this book weaves in more narratives of scientific inquiry and industrial innovation, to highlight aluminum's role in geoscience and engineering. I aim to capture the excitement of key technological discoveries that have come about from this metal and provide a fast-paced narrative rather than a deep historical exposition. Some of the questions to be explored are: Why are aluminum minerals so well bonded internally as to make extraction highly energy intensive? Despite its abundance in the earth's crust, why did life forms not evolve to metabolize this metal, given that they can consume so many others? How have debates about aluminum's toxicity taken root in alternative medicine circles? What led aluminum to be valued more than gold in the eighteenth and nineteenth centuries? Why is aluminum so recyclable compared to other metals, and how can we create a circular economy with this metal as a core infrastructure component?

The title of the book was carefully chosen to highlight the lessons that can be gleaned from nature regarding aluminum and its products. Aluminum ore is often soft and the consistency of *soil*, unlike the hard rock ores from which most other metals are extracted. Bauxite, aluminum's main ore, forms when laterite soils are severely leached of silica and other soluble materials in wet tropical or subtropical climates. Aluminum's relative ubiquity is thus deceptive and similar to the ostensible abundance of soils in general. The illusion of abundance can lead to many human follies. For example, the destruction by errant tillage practices in the American heartland led to the tragic displacement of over 3 million people during the Dust Bowl. On the other end of the product spectrum, aluminum *foil* exemplifies a material that is eminently recyclable but often not put through recycling streams because of the way we use it. Foil is thus an apt metaphor for human choices that determine how industry and nature can relate to each other. If we are to consider more sustainable material usage on the planet, then we need to think about how we design products for their ultimate use, reuse, and return to the resources cycle.

The narrative uses aluminum as a synecdoche for a broader conversation about planetary sustainability. What lessons can be learned from the material-energy nexus by considering the aluminum industry's path from extraction to recycling? How do mineral supply chains from developing countries to the developed world get monitored for improved social and environmental performance? What is the relationship between material discoveries and inventions? How can innovative technologies that emanate from raw material discoveries be functionally beneficial for society, not just drive ecologically deleterious consumer demand? Finally, the book engages with questions of ecological restoration. The abundance of aluminum in small quantities all over the earth's crust rather than in concentrated deposits has meant massive ecological landscape impacts. Based on fieldwork conducted over the past two decades at mine sites, I provide examples of restoration of the land and healing of communities affected by extraction. The mines-to-markets example of aluminum is a parable for the larger trials and triumphs of human material consumption. Such nuanced narratives around the challenges and opportunities of material consumption are now increasingly important as humanity struggles to address global environmental change in the Anthropocene.

Figure 0.1 clearly lays out how aluminum has among the highest cyclicality profiles of any metal and how, as with many other materials, China dominates the sector. China produced ten times as much aluminum metal in 2021 than the next country on the list. India, Russia, and Canada each produced comparable quantities (3.9, 2.7, and 3.1 million metric tons, respectively). Subsidized energy for smelting also put the United Arab Emirates in the top-ten list of producers, with 2.6 million tons. Clean geothermal energy in Iceland has also made it a favorable locale for smelting, and the country now produces as much aluminum metal as the United States (around 880,000 metric tons).[1]

Despite a range of synthetic materials coming online, a recent study titled *Opportunities for Aluminum in a Post-COVID Economy* suggests the metal's dominance is likely to last. According to this study, aluminum demand is forecast to grow by 33.3 Mt in the following decade, going from 86.2 Mt in 2020 to 119.5 Mt in 2030. More than 37 percent of this growth will come from China, a further 26 percent from the rest of Asia, 15 percent from North America, and 14 percent from Europe. The highest growth in terms of absolute demand will come from the transportation sector, which, driven by decarbonization policies and the shift from gasoline-burning vehicles to electric cars, will go from consuming 19.9 Mt of aluminum in 2020 to consuming 31.7 Mt in 2030. Most of this growth will come from China (33 percent), North America (22 percent), and Europe (19 percent).[2] Yet much of this demand could and should be met by recycling, given the high recyclability of the metal and the comparably very high energy cost of refining bauxite ore. According to the International Aluminum Institute, 75 percent of the 1.5 billion tons of the metal ever produced is still in use today, and over 30 million tons of aluminum scrap are recycled globally, making it among the most "circular" of all metal commodity supply chains. Yet the rates of aluminum recycling in the United States have largely plateaued, and there is still ample room for improving recovery of soda cans and foil. Such is the sparkling but salty story of this alluring metal that will unfold in these pages.

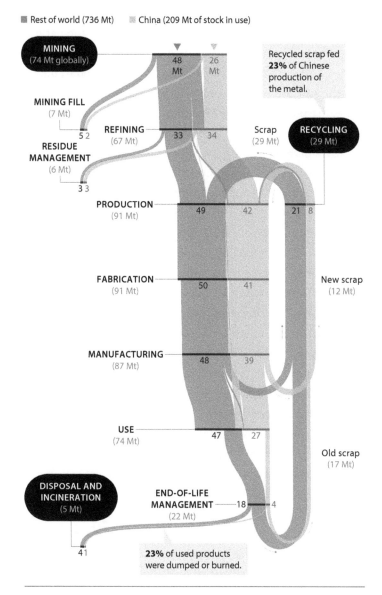

■ Rest of world (736 Mt)   ▨ China (209 Mt of stock in use)

MINING
(74 Mt globally)

48 Mt   26 Mt

MINING FILL
(7 Mt)

Recycled scrap fed
23% of Chinese
production of
the metal.

5 2

**RESIDUE
MANAGEMENT**
(6 Mt)

REFINING
(67 Mt)   33   34

Scrap
(29 Mt)

RECYCLING
(29 Mt)

3 3

PRODUCTION
(91 Mt)   49   42   21 8

FABRICATION
(91 Mt)   50   41

New scrap
(12 Mt)

MANUFACTURING
(87 Mt)   48   39

USE
(74 Mt)   47   27

Old scrap
(17 Mt)

**DISPOSAL AND
INCINERATION**
(5 Mt)

**END-OF-LIFE
MANAGEMENT**   18   4
(22 Mt)

4 1

23% of used products
were dumped or burned.

FIGURE 0.1   The global aluminum cycle.

Source: Y. Geng et al., "How to Globalize the Circular Economy," *Nature* 565,
no. 7738 (2019): 153–55.

The eight chapters are organized into four parts that consider the four stages of understanding the elements: (1) the natural occurrence of a resource and its place in natural systems, (2) primary source extraction and economic geology, (3) product development and use, and (4) cycling back from product use and restorative processes. Two chapters in each part are written in as narratives, with episodic segments that convey scientific content through storytelling. The quartet structure is also inspired by the four phases of aluminum's impact in human history and how it points a way toward planning for more sustainable resource management. In part 1, the fundamental geological and chemical properties of the element as it exists in nature are presented with geographic and historical illustrations of its occurrence and use. Part 2 depicts the historical moment when the metal was extracted from its salts and gained ascendance because of its unique properties. I examine the tenacity with which aluminum is bound in oxides and the story of how these bonds were broken. Part 3 lays out the phenomenal liberation of technology that the refinement of aluminum enabled, leading to an estimated hundred thousand different products by the 1950s. These products reflected the broader age of consumerism and the military-industrial complex that characterized much of the twentieth century. Part 4 comes full circle to the importance of cycles in resource flows and shows how aluminum products have been particularly successful in this regard. Finally, the epilogue presents the broader environmental-governance aspects of material flows that are being deliberated at the United Nations and other international bodies. This section is informed by my ongoing role serving on the UN International Resource Panel and recent efforts to establish a global treaty on mineral flows.

I am deeply grateful to Miranda Martin, my acquisitions editor at Columbia University Press, for sparking the idea for this book and shepherding the review process with rigor and efficiency. The three reviewers who went through the full manuscript and provided detailed feedback have helped improve the book. Such informed and constructive criticism is one of the main reasons I will always opt for university presses, even for a publication aimed at a broader audience. I want to share deep gratitude to numerous friends and colleagues with whom I had long conversations about the manuscript, in particular my former doctoral student Kim McRae, who introduced me to her contacts with the Mohawk communities that have been affected by the aluminum industries in upstate New York. Denise Humphrey Bebbington connected me with the filmmaker Esther Figueroa, whose insights were immensely helpful in understanding the field perspective on bauxite extraction in the Caribbean.

The staff members of the Oberlin College library, led by the archivist Ken Grosso, where the archives of Charles Hall are kept, were also very welcoming and helpful despite COVID protocols at the height of the pandemic. I am also deeply appreciative of my affiliation as honorary professor at the Sustainable Minerals Institute at the University of Queensland, Australia, where I have learned so much from colleagues about ways of connecting our needs for minerals with broader questions of sustainability. In particular Deanna Kemp, Lynda Lawson, and Glen Corder provided valuable insights and case examples for this book. Lynda also introduced me to the work of Penda Diallo on the political economy of bauxite in Guinea. I am especially grateful to Penda for introducing me to their contacts in Guinea to better understand this vital case study. Special thanks also to Nellie Mutemeri, with whom I have served on the Independent Governance Committee of the Dubai Multicommodity Centre (DMCC) and who also facilitated valuable contacts. I first met Fiona Solomon during my time in Australia, and her valuable input as the chief executive officer of the Aluminum Stewardship Initiative provided this book with a hopeful coda. My family in the United States, Canada, and Pakistan were supportive as always in allowing me the time and space for writing. My research assistant, Nabeel Trimzi, patiently worked through assembling the diagrams and helped with bibliographic research. I am grateful for his diligence and positive engagement with whatever tasks I sent his way.

Finally, this book is dedicated to Larry Susskind, who mentored me in my earliest days of research activity more than twenty years ago. I was then the youngest doctoral student in his advisory cohort, with little direct research experience, and I felt daunted by the excellence around me at the Department of Urban Studies and Planning at MIT. He found a way for me to connect my roots in chemistry with broader social questions about how we source minerals. Larry "thinks big" but has a practical, down-to-earth approach to problem solving. He inculcated a balance between theoretical insight and applied pragmatism in research, which has been my guiding light throughout my career and hopefully shines through in this book as well.

SOIL TO FOIL

I

SALT AND SOD

You are the salt of the earth.
But remember that salt is useful when in association,
but useless in isolation.

—Israelmore Ayivor

1

ELEMENTAL ORIGINS AND THE INVENTION OF NEED

Chemistry itself knows altogether too well that—given the real fear that the scarcity of global resources and energy might threaten the unity of mankind—chemistry is in a position to make a contribution towards securing a true peace on earth.

—Kenichi Fukui

Off the windswept western coast of Florida, a slender island called Sanibel has particular appeal to chemists. It is not the salt water of the Gulf of Mexico or the sand of its beaches that attracts their interest but the memories of gatherings gone by. In the 1960s, during the height of the Cold War, a symposium was hosted there to bring together from all over the world for intimate discussions bright minds in quantum chemistry, biology, and physics. The sunny weather and quiet environs were intended as creative stimulants. In the winter of 1964, Kenichi Fukui, a chemistry professor from Japan's Kyoto University, met at Sanibel Island for the first time a young doctoral graduate from Harvard named Roald Hoffmann. They were both working on similar research areas and struck up a friendship that would last a lifetime.

A shared desire to make their esoteric work relevant to the world's challenges of resource scarcity strengthened their bond beyond the usual academic exchanges. They would go on to share a Nobel Prize in chemistry in 1981 for pioneering work that explained the speed and trajectory of chemical reactions. Hoffmann would later develop his career at Cornell University in upstate New York as a chemistry professor but also as a poet, a playwright, and for many years the host of a weekly café cabaret called "Entertaining Science" in Manhattan.[1] He would go on to develop twenty-six episodes of a pioneering educational series for public television called *The World of Chemistry*. Throughout his career, though, Hoffmann's friendship with Fukui was a constant. The two scientists had connected on earthly matters in a palpable way: through a passion for pottery. They could sit for hours together watching masters of ceramics in Kyoto wielding their art to convert soil into splendid vessels.

Coming from an island nation with limited natural resources, Fukui had observed how scarcity had driven Japan's expansionary impulses and ravenous industrial development. He had lived through the horrors of World War II and knew all too well the roles science, chemistry, and raw metallic elements had played in the bombing of Hiroshima and Nagasaki. This devoted imperial citizen had also worked in the Army Fuel Laboratory during the war, when there was little freedom for scientists to choose their careers. Amid the chaos of the times, he still found quiet moments to reflect on how quantum chemistry could serve to solve many of the problems that its industrial applications had created.

As Japan's first Nobel laureate in chemistry, a heavy burden fell on him to use the pulpit of the prize for public education and broader interaction with policy developments in postwar Japan. In the country's new constitution, science was to be reconfigured toward goals of human development. In the words of the historian John Dower, Japan remarkably showed the world how to "embrace defeat." The scarcity or abundance of resources was no longer defined by parochial measures of what could be domestically harnessed at high ecological costs but rather through an understanding of natural cycles and resource efficiency. Thanks to the work of scientists like Fukui—a true son of his soil—Japan became the most ecologically efficient advanced industrialized country. Despite a xenophobic ethos and repugnance toward immigration, the Japanese scientific community sought a planetary vision to problem solving. The Land of the Rising Sun—the only nation to have endured a full-scale nuclear bombing—became one of the largest donors to the United Nations Environment Programme.

Hoffmann and Fukui's friendship in science had depth because it had a panoramic planetary vision. Their camaraderie was cemented by a deep knowledge of and personal exposure to the materiality of war and peace. Fukui had endured the bombing of Hiroshima and Nagasaki, and Hoffmann had survived the horrors of the Holocaust in what is now western Ukraine. His father perished in a concentration camp, and both he and his mother had to hide in the attic of a church house for almost a year before escaping to America. It was a friendship built around a common experience of trauma in early life that gave them perspective on our place in the universe.

In an online video interview with Hoffmann in early 2021, I was warmly treated to a collection of joyful memories. The office backdrop had an assemblage of travel art and mementos and a graying portrait of Fukui, which Hoffmann noted gave him "a feeling of peace and inspiration." Upon Fukui's death, Hoffmann wrote his obituary in the journal *Nature* and stated that his co-laureate was "driven by the truest desire to understand the world."

Hoffmann also shared with me an article that Fukui had written on the "Uniqueness of Nature and Human Beings." It was based on a speech given at the Nobel Jubilee meetings in Uppsala, Sweden, and had been published in the *Journal of Quantum Chemistry* after a grueling two-year peer-review process in 1994.[2] In this article Fukui wonders over the "sense of alienation and isolation felt by humans at the emergence of numerous materials and phenomena not found in Nature but created by the enormous power of science and technology." This wonderment and anxiety over the creative powers of Nature versus those of humanity were important to Hoffmann and Fukui. They knew all too well that such existential challenges for humanity could only be solved if we considered the constraints imposed on our resource use by the elements of the earth. The need to steer our material culture and our technological advancement between war and peace— and between need and greed—was what Japan, America, and indeed all nations learned during the second half of the twentieth century. Navigating the boundaries between need and greed at the individual and collective levels remains the ultimate ecological challenge for humanity. To gain a more profound understanding about the elements that animate our lives, we must first venture back to our most fundamental material origins. With humility, let us consider our diminutive place in the cosmos. From this narrative journey deep in time, we will find an element that might be a suitable parable for our evolving relationship with nature.

NUCLEAR BIRTH OF THE ELEMENTS

The first three minutes of the universe's existence largely produced only four elements—hydrogen, helium, and fractional amounts of lithium and beryllium. This quartet of elements then engaged in the nuclear reactions that eventually gave us an entire periodic table of material building blocks. The explosive existential energy burst that we commonly call the Big Bang is widely accepted as inaugurating our current state of cosmic reality. Our material realm was to evolve from the conversion of the energy of that stupendous Big Bang into mass. The infant universe was able to make this conversion occur fairly quickly, in conformity with Einstein's celebrated equation $E=mc^2$. Those first three minutes had such creative consequence that the physicist Steven Weinberg authored an entire book by that title. It was a short but momentous period in the universe's history. Extremely high temperature during those three minutes allowed for energy-to-matter conversion for the simplest of elements. Most of the hydrogen that will ever exist in the universe was formed during that period. As the cosmos cooled, the formation of other elements was accomplished through other mechanisms, because nuclear fusion reactions could not be sustained at lower temperatures.[3]

The initial froth of embryonic elements was a messy mix of gaseous particles. For several thousand years after the Big Bang, the elements were merely naked, positively charged nuclei without massless electrons orbiting around them. As the cosmos cooled further, a natural clumping of matter through gravitational attraction occurred. During this juvenile period of our universe, massive stars formed via these gravitational mechanisms in the spiraling arms of swirling galactic material. The stars followed physical laws of emergence across space-time and became the nursery where many other elements would be born. The primordial elements—hydrogen and helium—were the fuel for nuclear fusion reactions, which added protons to atoms, forming new elements. Elemental identity became defined by the number of protons in an atom.

Stars would take varied trajectories of size, temperature, and density, which would determine their fate and their facility for elemental incubation. Stellar cores could sustain creation up to the elemental metal iron, which has twenty-six protons; after this critical threshold, stars become unstable and begin to collapse. Some stars would explode stupendously in supernovas, which would allow for larger elements to be synthesized in the energy of the stellar cataclysm. Other stars would become dominated by chargeless neutrons and find themselves falling into gravitational death spirals with their neighbors or collapsing in upon themselves to form neutron stars and black holes. Colossal collisions of neutron stars produced further nuclear reactions that created yet more elements.

The discovery of these stellar processes of elemental synthesis proved to be so important for understanding contemporary material life that their discoverers got their initials enshrined on the common title of a paper for posterity. The "B2FH paper," published by the *Review of Modern Physics* in 1957 and formally titled "The Synthesis of Elements in Stars," has its own Wikipedia page! The lead author was the British astronomer Margaret Burbidge, followed by Geoffrey Burbidge (her husband), William Fowler, and Fred Hoyle. Appreciating the momentous nature of their topic, the authors open the 103-page paper with two contending quotes from Shakespeare—highly unusual for an esoteric physics journal:

> It is the stars; the stars above us, govern our conditions.
>
> (*King Lear*, 4.3)

> The fault, dear Brutus, is not in our stars, But in ourselves.
>
> (*Julius Caesar*, 1.2)

Before the research presented in this paper, the widely accepted view was that all elements were formed during the Big Bang. However, the Burbidges had always been skeptical about subscribing to the theory of the Big Bang and sought other explanations for elemental origins. They were joined in their skepticism by Sir Fred Hoyle, who had coined the term "Big Bang" in an impromptu episode on BBC Radio in 1949 but then dismissed the theory. Instead, Hoyle and the Burgesses embraced an alternative theory of origin called the "steady-state model," whereby matter might have been created uniformly across the universe. Their refusal to accept the Big Bang orthodoxy after the 1960s, when observational evidence began to suggest its likelihood, may well have deprived them of a Nobel Prize for their seminal work on nucleosynthesis.

Meanwhile, the fourth author of the paper, William Fowler, did go on to become a Nobel laureate in physics in 1983 "for his theoretical and experimental studies of the nuclear reactions of importance in the formation of the chemical elements in the universe." Even though much of the observational labor was done by Margaret Burbidge in authoring this definitive paper, she was not directly recognized with this accolade. She did most of the research on the paper while pregnant and in defiance of the repugnant stereotypes of female underperformance in science. Over the course of her 101-year-long life (1919 to 2020), she did win many other awards, including the National Medal of Science, from President Reagan in 1983, and the Gold Medal of the Royal Astronomical Society.

Despite his brilliance and knighthood, Hoyle distanced himself from the mainstream of science by rejecting the empirically dominant view of the chemical origins of life on Earth. Instead, he subscribed to the fringe theory of "panspermia," which held that life was brought to Earth from outer space via dust, spores, and meteorites. Toward the ends of their lives, the Burbidges and Hoyle suggested a hybrid model for universal origin whereby "mini Big Bangs" may occur within a steady-state model. Yet this too did not find much currency, and the trio today is mostly remembered through the acronym on their journal article on the stellar origin of the elements.

While we owe our elemental existence to stars, as revealed in B2FH, these celestial luminaries have only a trivial effect on our lives and destinies, with one singular exception—the sun that shines daily on us. Thus, King Lear and Julius Caesar were both right—depending on the context and scope of ascribing credit for our fortunes. Each day of our being alive on this habitable planet we owe to this star and its energy. We seek to harness solar power as the ultimate "renewable" source of energy. The astute (or captious) astronomer would note, however, that the nuclear fusion reaction that is converting hydrogen to helium and other elements in our sun is not reversible within the solar lifespan of a few billion years—that is, the sun will eventually burn out. But renewability and cycles are, of course, a function of what timescale we choose for our analysis and how we consider the stability of chemical compounds. On human timescales, the sun offers the paradigmatic unlimited resource.

The stability of each element, as we understand particle physics thus far, has been determined by a tussle between the four fundamental forces of nature: the electromagnetic force, the weak and strong nuclear forces, and gravity. Adding an additional neutron to a nucleus can add stability or lead to an inexorable release of radioactive energy and the potential decay of that element into another. As the number of protons increases in nuclei, the chance for instability also increases, but nature can still find what nuclear physicists call "islands of stability" for many of the higher elements as well. The inextricable relationship between mass and energy is a constant quest for stable states resulting from a complex tussle among these fundamental forces of nature. More than 14 billion years after the Big Bang, on a small speck of a planet, a complex organic assemblage of molecules with agency calling itself *Homo sapiens* would figure out a way to tentatively tame these forces.

The alchemists' dream of converting lead to gold or other elemental transformations was realized frequently in the early universe through the power of nuclear fusion. Physics gives way to chemistry in operational science once elements form and begin to engage in reactions and bonding between themselves. These reactions can produce molecules of increasing levels of complexity. Once these molecules become large and complex enough, they can gain fantastic properties of replication, folding, fracturing, and recombination. Like a vine on a trellis, these molecules find the paths of least energetic resistance and weave a chemical tapestry. The structural edifice of natural laws forms the trellis on which these molecular marvels take root. Once we are beyond the specter of engaging in nuclear reactions because of energy limitations, each molecule is constrained by the elements that it can work with in creating the infrastructure of life.

The abundance of these elements in the universe is highly variable and not directly related to the sizes of their nuclei. Instead, it is determined by complex criteria for nuclear stability that are roughly instantiated by an even and odd numbering of protons. In general, atomic even numbers for elements have greater abundance as a result of nuclear stability. However, elemental order in this context also has many exceptions, as shown in figure 1.1. The saw-tooth curve of elemental abundance in the universe identifies anomalies like beryllium, which has an even atomic number but is very scarce. Because of the small size of its nucleus, the fusion of two helium atoms to form beryllium

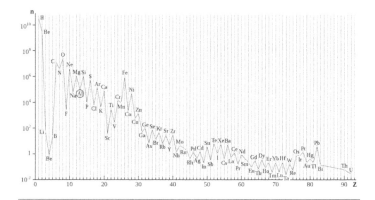

FIGURE 1.1 Chemical abundance of the elements in the universe, where Z = atomic size (chemical abbreviations for each element in appendix A; Aluminum [Al] is circled).

Source: Wikimedia Commons.

creates an unstable isotope (a form of the element with a different number of neutrons). Thus, stable beryllium, which we find in emeralds on Earth, was formed by cosmic rays splitting larger nuclei, a rare event. Lithium and boron owe their scarcity to the same challenges of formation. Fluorine gives a sudden dip in the curve, more so than other elements, because of cycles of nuclear burning in which it has a propensity for getting involved.

The vertical axis of figure 1.1 uses a logarithmic scale (shown in exponential format); each unit represents a tenfold change. This scaling is necessary because, given the vast variation in abundance among the elements, this is the only way to fit all the points meaningfully on a chart. Hydrogen is normalized for abundance by the number twelve, out of convention for the number of months in a year. Yet the abundance of elements in the universe is not necessarily correlated with the abundance of these elements on Earth. Because of the ease of its formation, hydrogen constitutes almost 75 percent of the observable mass of the universe; helium is around 24 percent. All the other elements comprise the remaining 1 percent of the observable mass. The elusive category of "dark matter" also shows up indirectly through gravitational and astrophysical measurements but has not been directly observed.

Alarmingly, the measurements suggest that dark matter could constitute 85 percent of the mass of the universe. Dark matter should of course not be confused with "antimatter," which is observable (and indeed used in medical devices such as PET scanners) but highly scarce in the universe. The reason for the scarcity of antimatter remains a major mystery for physics. Mathematical models of the Big Bang suggest that equal measures of both matter and antimatter should have been produced at the time of the singularity of existence. Matter's triumph over antimatter essentially allowed reality to exist: these two elemental forms mutually disintegrate into other minor particles when brought into proximity.

The material evolution of the universe through multitudes of stellar and galactic cycles created new clusters of elemental order. Material convergence within the void of space began to occur in specific clustering regions simply called galactic "walls." These walls followed emergent patterns of order, often motivated by numerous interacting forces, like veins of a crystal lattice forming on a windshield on a cold wintery night. Spiraling agglomerations of material called galaxies emerged along these walls. Nested clusters of elemental dust gave way to nebulae that in turn gave way to stars and their orbiting progeny of material.

Gravity began to dominate as a force in solar systems. Mass slowly accumulated through gravitational attraction and opportune collisions between cooling materials to form solid-mass planets. In other cases, gases cohered into voluminous planetary masses that did not quite have the size or density to start nuclear reactions and become stars. Instead, these entities became the gas giants, like Jupiter, which is not so much a gargantuan planet as it is a failed star. The elements that formed the planets also diverged in their prevalence or scarcity. Hydrogen and helium lost their hegemony of abundance in planetary systems, and other elements that could form tighter bonds and geological compounds gained ascendance. Many of those elements, which we now call the base metals, like iron, benefited from gravity, as their denser binding gave them a greater ability to concentrate in planetary cores. But not all metals followed that path to the center. Some loitered on the surface of these newly forming rocky planets, where, at least on Earth, much of the drama of life was to unfold.

THE PRIMACY OF ELEMENT 13

The early Earth was a product of multiple collisions of rocky debris within the solar system. In addition to the collisions of asteroids and comets, Earth also endured some cataclysmic collisions with larger objects. Most notably, the formation of the moon is believed to have been the result of such a collision with a planetoid about the size of Mars, which also gave our planet its particular elemental composition. The original composition of Earth is believed to have been similar to that of many of the celestial masses hitting Earth. As the repeated hits deposited more mass, the planet reached a critical diameter of 320 kilometers of rock, with a mass great enough to generate enough heat to melt the rock at its center—what we call magma. Given Earth's proximity to the sun, the meteors hitting the young planet had to be able to withstand solar heat as well without being vaporized and dissipated. Hence metallic materials, with their high melting points, became particularly privileged planetary building blocks. Among the multitudinous metals in this collection, the element with atomic number 13—aluminum—had an intriguing rise to dominance in Earth's crust. Rather than being doomed to oblivion by its ominous numerology, element 13 flourishes in Earth's upper crust.

Eighty-five percent of the meteors that are able to reach Earth are of a rock form called chondrite. This metalliferous rock contains the solidified remains of molten mineral droplets surrounded by calcium-aluminum inclusions. If you slice open a sample of this brand of meteorite in any major natural history museum, you will find the cross-sectional evidence of that tumultuous moment of Earth's past: molten minerals and lighter metals encrusted in a hard amorphous lump, crashing onto a seething orb of magma. The aluminum in these chondrites also gives us a way to estimate Earth's age through a range of measurements of radioactive decay of the element's isotopes and its conversion to magnesium.

Tracing our way back through these calculations, we learn that Earth's mass, as we know it, is around 4.6 billion years old. The early Earth was a hellish place—quite literally a land of fire and brimstone. That geological eon from 4.5 to 4 billion years ago is aptly called the Hadean, after Hades, the Greek god of the underworld. The venerable earth scientist Robert Hazen has posited that minerals have also followed an "evolution" through Earth's history and have slowly become ever more complex through the various eons after the Hadean. As with biological evolution, various "species" of minerals differentiated and were organized via thermodynamic pathways. Furthermore, Hazen calculates that more than half of the Earth's 4,400 or so mineral types were formed through interactions with living cellular forms. For that reason, particular minerals began to dominate the Earth's crust where conditions were favorable for life, and these manifested in highly diverse physical forms.[4]

The Hadean eon saw the fundamental organization of the elements across the planet. This was a time when rocks we now consider nonrenewable on human timescales were cycling through the planet in convection currents via constant volcanism. As the planet cooled, the elements further differentiated themselves based on geochemical affinities. The great Norwegian mineralogist Victor Goldschmidt observed how various elements adhered to one another and formed the vast edifice of the Earth's crust. Instead of our usual classification of the periodic table into metals, nonmetals, and metalloids, based on the physical properties the elements display, Goldschmidt offered a geochemical delineation of the elements into four categories: (1) lithophile (rock-loving and largely forming oxygen compounds), (2) siderophile (iron-loving and forming bonds therein), (3) chalcophile (ore-loving and largely forming bonds with sulfur, selenium, and tellurium), and (4) atmophile (gas-loving and largely remaining as such at room temperature).

Figure 1.2 shows the salience of element 13 in this lexicon and in Earth's crust. Essentially, aluminum was an extreme lithophile, and owing to its specific electronic configuration, it formed very strong bonds with oxygen. Element 13's love affair with oxygen led it to accumulate in Earth's crust despite its relative rarity in the universe. While other metals like iron and nickel precipitated their way down to the core, aluminum bonded with oxygen to form light rocks that were able to spread diffusely across Earth's crust.

Aluminum is uniquely positioned on the periodic table in the right "armpit," between the terminus of transition metals in the middle and the twilight zone of "metalloids." Although

	1	2	3	4	5	6	7	8	9	10	11	12	13	14	15	16	17	18
1	1 H																	2 He
2	3 Li	4 Be											5 B	6 C	7 N	8 O	9 F	10 Ne
3	11 Na	12 Mg											13 Al	14 Si	15 P	16 S	17 Cl	18 Ar
4	19 K	20 Ca	21 Sc	22 Ti	23 V	24 Cr	25 Mn	26 Fe	27 Co	28 Ni	29 Cu	30 Zn	31 Ga	32 Ge	33 As	34 Se	35 Br	36 Kr
5	37 Rb	38 Sr	39 Y	40 Zr	41 Nb	42 Mo	43 Tc	44 Ru	45 Rh	46 Pd	47 Ag	48 Cd	49 In	50 Sn	51 Sb	52 Te	53 I	54 Xe
6	55 Cs	56 Ba		72 Hf	73 Ta	74 W	75 Re	76 Os	77 Ir	78 Pt	79 Au	80 Hg	81 Tl	82 Pb	83 Bi	84 Po	85 At	86 Rn
7	87 Fr	88 Ra		104 Rf	105 Db	106 Sg	107 Bh	108 Hs	109 Mt	110 Ds	111 Rg	112 Cn	113 Nh	114 Fl	115 Mc	116 Lv	117 Ts	118 Og

57 La	58 Ce	59 Pr	60 Nd	61 Pm	62 Sm	63 Eu	64 Gd	65 Tb	66 Dy	67 Ho	68 Er	69 Tm	70 Yb	71 Lu
89 Ac	90 Th	91 Pa	92 U	93 Np	94 Pu	95 Am	96 Cm	97 Bk	98 Cf	99 Es	100 Fm	101 Md	102 No	103 Lr

FIGURE 1.2 Situating element 13 in the periodic table.

Source: Wikimedia Commons.

chemists detest the categorization of elements in terms of metals, nonmetals, and the in-between category of "metalloids," there is functional value to such terms. The metalloids exhibit ambivalent and hard-to-categorize characteristics in terms of the chemical reactions they undergo and the physical properties they display. The six most widely accepted metalloids are boron, silicon, germanium, arsenic, antimony, and tellurium. Aluminum has also been occasionally classified as a metalloid because of its complex bonding behavior. Its neighboring element silicon is a rival for oxygen's attraction.

Silicon oxides constitute common sand and glass. They are the most ubiquitous of all materials in Earth's crust, and aluminum has competed with silicon in forming the most stable of planetary rocks. Occasionally, there is also a threesome attachment between these excitable elements—the result is aluminosilicate rocks, key constituents of common clay. More than two-thirds of Earth's crust by mass comprises these three elements—oxygen (47 percent), silicon (28 percent), and aluminum (8 percent), which are often closely bonded with one another, forming stable slabs of minerals.

Abundance is a relative term in earthly matters. Although element 13 is the most abundant metal in Earth's upper crust, its concentration across the crust is highly diffuse, and it does not occur in solid elemental form naturally like gold or copper. There is a story from the Roman courtier Gaius Petronius, in his *Satyricon*, in which a glass goblet is gifted to the emperor that does not break when thrown on the floor but instead deforms. Might this goblet have been made from a fine sheet of aluminum? Highly unlikely. Aluminum metal was not isolated from its compounds until several centuries later, and the metal has never been found in natural metallic form anywhere on Earth. In recent years some artifacts made of aluminum were also claimed to be from the Jin dynasty in China (265–420 CE), which would have suggested the availability of the metal either through natural deposits or the existence of refining capabilities during this period. However, detailed chemical analysis revealed these to be fakes either deliberately or accidentally inserted into tombs excavated in the 1950s.[5]

Despite its lack of virgin availability on Earth as a solitary element, aluminum's compounds spread their splendors across a vast array of rocks. The weathered rocks then find their way through rivers and underground springs and get deposited around the planet. Aluminum's ubiquity in crustal rocks has meant that its deposits also are varied. From desert salt pans to tropical islands, aluminum compounds have accumulated in sizeable deposits and in highly diffuse physical forms. Oxides of aluminum can consist of a soft crumbly agglomeration of sedimentary rock called bauxite or crystallize as precious corundum crystals that we commonly call rubies and sapphires. Aluminum can also combine with other reactive metals like sodium and potassium to form complex salts that churn through volcanic processes and deposit themselves around hot springs and geysers. It was in such environs that humanity first found the functional form of aluminum and from which its name was eventually derived.

PURPLE PURPOSE

Among the most enchanting natural vistas on the planet is the Grand Prismatic Spring at Yellowstone National Park in the eastern reaches of America's Rocky Mountains. An aerial image of the hot spring shows a dramatic blue interior where the water is hot, deep, and devoid of any life. Surrounding the brilliant blue eye of the pool is a chromatic profusion of rusts, greens, yellows, and even an occasional purple. The varied colors of the prismatic spring at the periphery are literally a gift of life. Each hue is caused by a different microorganism secreting colorful chemicals into the environment. The natural world is animated by color. It gives our eyes the ability to differentiate substances and lets us ascribe feelings and meaning to what we observe. Color has no physical reality on its own. It is merely a label for a wavelength of light passing through some medium that reaches our retina and is registered by our brain as having a particular conscious, subjective quality. Elements often imbue color based on the molecular structure that they cause in a crystal. Living things have evolved the ability to use the elements to craft colorful compounds that give their exteriors a selective advantage in evolution. Humanity has admired the colors produced by these organisms, and we have adorned our material possessions with nature's palette.

The human desire to appropriate color is most manifest in what we wear. We were not content just to have colors embellish the walls of our habitats; we needed to have color closer to our bodies. The desire to adorn ourselves with cosmetics and colorful cloaks may well have inspired humanity's first chemical industry. Natural colors derived from plants or animal secretions are attractive but often ephemeral when used as dyes. They wash off with water and wear or simply bleach in the sun. To give them permanence on fabric or leather, further chemical interventions are needed. Many ancient civilizations experimented with a range of earthly materials that were added to dyeing vats and tested on fabrics. These compounds are now referred to as "mordants"—after the Greek word for "bite." Among the most potent mordants is an assemblage of slippery salts found in evaporated lake beds or near volcanic and geothermal features.

The ancient world saw the value in the "bitter salts"—*alumen* in Latin—that laced these locations. Many centuries later, Saint Isidore would playfully ascribe the origins of the word also to "a-lumen"—"so called 'from light' [lumen] since it gives lightness to the dyeing tints." One of Aristotle's pupils in science, the long-lived Theophrastus (372–287 BCE), wrote a treatise titled *On Salts, Soda, and Alum*, which was lost in antiquity but documented by the historian Diogenes. This text is believed to be the earliest scientific examination of specific compounds of element 13—which at the time was of course not isolated as such. Only a few fragments of papyrus of this manuscript have been found and analyzed, and they suggest a breadth of uses of alum in ancient textiles. These alums were noted to hold particularly powerful properties as mordants and were also described by Pliny the Elder in his seminal *Natural History* a couple of centuries later. In lyrical terms Pliny would note: "Alumen is saltness [salsugo] of the earth formed in winter from water and clay; and matured in summer from the sun's rays." Pliny further suggests another macabre use of this peculiar stone, found in the territory of Assos, to hasten the decomposition of cadavers. Alum was cast as a "flesh-eating" stone and was enshrined in the sarcophagus industry (*sarksarc* [flesh] + *phagein* [to eat]).

Corpses buried with alum were noted by Pliny to decompose in forty days. Famed for this property, the sarcophagi produced from the andesite in Assos were in high demand in the ancient world and exported to Lebanon, Syria, Greece, and Rome. Alum's elemental properties seemed also to be in synch with Aristotle's view of "emanations" and a recognition that there was some primordial element that gave this substance special properties. Aristotle saw alum as a manifestation of earthly processes. In his book *Meteorologica*, he noted that the presence of alum (also called stypteria) was a residual material from "flavors" of water.[6]

Among the textiles for which alum was particularly important were purple robes used by royalty. The color purple became associated with imperial greatness because of its rarity and brilliance. Purple dyes in nature were few and far between. Plants that produced purple flowers had ephemeral pigments that lost color over time. A remarkable source of purple dye was extracted from the glands of a sea snail species in the Mediterranean. Unlike plant pigments, this purple mucus from the snail did not fade in sunlight and in fact became more pronounced in coloration over time. The coastal residents of Tyre in present-day Lebanon became major harvesters of several species of the *Murex* sea snail to extract this precious purple dye. To this day, the color of this dye is known as Tyrian purple. The predatory snails produced the dye as a defense mechanism and could be stimulated to do so while alive in a controlled setting. However, such a "milking" of snails for the dye without killing them was labor intensive, so instead thousands of snails were cooked in vats to release the dye. The process may well have been devised by the Phoenicians, as the primacy of purple can be traced back to around 1500 BCE. Etymologists have even suggested that Phoenicia means "land of purple," owing to this marvelous biotic discovery that gave them such a lucrative and colorful trade.

Yet trade in the violet splendor of the *Murex* snail's mucus needed to be coupled with the availability of alum salts as mordants. The aristocracies across the Mediterranean demanded alum to meet their desires for purple prestige. Alum trade routes developed across the region. Egypt had some of the most ancient sources of alum from desert oases such as Dakhla and Kharga, where ancient lakes had evaporated to leave a range of deposits. These alum mines went back to at least 2000 BCE, when the pharaohs were believed to have used the material for colored glass glazes. The historian Herodotus noted that Egypt's last indigenous pharaoh, Amasis, was a devotee of the oracles of Delphi and in his contribution to the construction of a new temple gave a gift of a thousand talents' weight of alum.

Small islands in the Mediterranean also became major alum powers. Aristotle had conducted fieldwork on the islands of Lesbos and Milos and observed alum deposits there near volcanic vents. Further west and just north of Sicily in the seismic shadow of the great Mount Etna, an assemblage of cratered islands also became major alum-mining regions. Of particular appeal was a small dormant island called "Vulcano"—named for the Roman god of fire—and the root of our contemporary word "volcano." There are remains of alum-processing sites around the island's calderas and a famed "alum cave" that still boasts impressive crystals.

A perfect crystal of alum from a slightly acidic solution has eight faces, all of equal equilateral triangles—such an octahedron is one of the regular Platonic solids. Aluminum minerals that constitute a variety of "alums" are known to crystallize rapidly depending on the solvents and seeding agents used to stimulate the process of molecular structuring. The procurement and processing of alum for use in dyes is believed to be the "earliest chemical industry," which provided the title of a monumental history on the material by the historian Charles Singer, published in 1948. Alum was thus among the earliest chemicals with which humans experimented. Biological fluids were often a key part of the experimenting process. If a rock containing aluminum sulfate is boiled with decomposing urine, you get ammonium alum—which was found to have remarkable fire-retardant properties. Purple robes for the elite that had been soaked further in alum solutions were thus even more prized. Without knowing the intricacies of the chemistry of element 13, the ancients indirectly made aluminum a prized commodity. What started as a lust for rare, robust color and safe attire slowly opened the doors to the discovery of an elusive element in its metallic form.

CHEMICAL COLLUSION AND THE CHURCH

With the rise of empires in Europe, the demand for formal colorful clothing grew rapidly. Not only were wealthy merchants and the aristocracy coveting alum, but with the ascendance of papal Christianity, the Catholic Church also began to take a strong interest in the alum trade. In addition to its use as a mordant, alum's medicinal properties also gained greater attention and currency during the medieval period. Although the medicinal properties of alum had been championed in antiquity, the Islamic alchemist Razi (865–925) was perhaps the first to empirically demonstrate certain properties of the material. In his book *Secret of Secrets* Razi describes the establishment of a laboratory, and he listed alum among his minerals of medicinal importance. Farther east, the Chinese had also independently discovered the mordant and medicinal properties of alum. They called minerals associated with alum *pei fan* and used the material as an astringent, detersive, and styptic agent. Some of these uses were known as far back as Alexander's time, when the young conqueror returned from Asia, bringing with him colored silk laced with alum. The Islamic scholars also used many of the Greek texts in their own scientific research. The uses of alum thus exemplified a remarkable cross-current of chemical learning along medieval trade routes.

Christianity's association with alum became particularly noticeable during the Byzantine period, when alum mines in northern Anatolia gained concessions for trade by Genoese merchants in 1275. There are more than one thousand recorded underground alum mining sites in what was then also referred to as Asia Minor. Alum processing as a mordant had been perfected farther east as well by Muslim carpet makers in Persia. In the 1330s, the Berber travel writer and scholar Ibn Battuta noted in his writings a place called Aksaray, near the city of Konya. This was the region famed for the thirteenth-century saint Rumi and was a center of textile art and culture. The use of wool for carpet manufacturing also required a process called "fulling," which involved the curing of the wool with alum. By the time of the Crusades, in the late eleventh century, fulling mills were active throughout the medieval Islamic world, from Islamic Spain and North Africa in the west to Central Asia in the east. The alum industry became yet another economic incentive for invasion and conquest during the Crusades. Vannoccio Biringuccio in his 1540 book *De la pirotechnia* noted that alum "is no less necessary to dyers of wool and woolen-cloth than bread is to humankind."

The Genovese dominance of the alum trade from Anatolia ended with the fall of Constantinople (present-day Istanbul) to the Ottoman Muslim Turks in 1453. This led to a massive scramble to find alternative sources of alum within Christendom. By 1458, the European stocks of alum had been exhausted, but then a few new deposits of alum were discovered at Volterra in Tuscany and Agnano in Naples. Even more serendipitously for the Catholic Church, in 1462 rich alum deposits were discovered in Mount Tolfa, near Civitavecchia, which fell within the Papal States. The feudal owners of Tolfa mediated a deal between Pope Pius II and the affluent Medici family of bankers in Florence to create a long-term contract that essentially saved the Holy See from insolvency. The Societas Alumnum was essentially a cartel arrangement between the Medici bank and the church. Upon entering the papal cartel in 1466, the Medici family set out to definitively eliminate all competition. They first suppressed production of the mines at Volterra, which was under Florentine rule. When the citizens of that region objected to the loss of income and threatened to break away from Florence, Lorenzo de Medici used his political influence to quell the revolt, which was accomplished by military siege. Next, the papacy joined the Medicis' attempt to eliminate Turkish competition by banning all imports. These actions by the Medicis in the alum trade produced a significant portion of their wealth and consequent patronage of the arts and sciences during the Renaissance. Thus we can indirectly credit the alum trade for giving us the magnificence of Michelangelo's frescoes or the blossoming of Leonardo da Vinci's genius.

The terms of the papal contract with the Medicis, executed in 1466, were as follows: The Medici became partners in the Societas

Aluminum and took over operation of the mines and the sale of its entire output. They agreed to pay a royalty of two ducats to the Apostolic Camera (essentially, the papacy's treasury) for each *cantaro* of alum taken from the papal warehouses at Civitavecchia. All the alum mined at Tolfa was to be stored in those warehouses in order to prevent any leakage to the detriment of the papal treasury, and nothing could be taken from the warehouses except in the presence of a papal official. Finally, in addition to the royalty, the Apostolic Camera was to receive two-thirds of all extra profit. The pope earmarked the income from the Tolfa mines for the crusade against the Turks and the Hussite heretics. He also forbade, under threat of ecclesiastical censure, the importation of Turkish alum into Christian territory.[7]

Interestingly enough, a few centuries later attempts at forming a cartel for aluminum metal through the International Bauxite Association failed, and the metal began to be traded on the London Metals Exchange. The success of cartels is often dependent on a political unity of action and the mitigation of violations. This was possible in the papal days of alum holdings but less so with the sources of the metal in the twentieth century.

Between 1462 and 1796, some 17 million tons of alunite, from which alum was extracted, were taken from the papal alum mines at Tolfa. Techniques for the production of alum from alunite, which changed little over this period, involved four stages that served to remove impurities and render the substance into a commercially viable form. The four stages were calcination (roasting in a kiln for ten or more hours), maceration (sprinkling heaps of calcinated rock with water once or twice a day for forty days, causing it to separate), lessivage (boiling the resulting material to separate impurities), and crystallization. In 1471, Pope Paul I estimated that the annual revenue of the church was around 200,000 florins. The alum trade is believed to have brought in an additional 100,000 florins annually, thus accounting for a third of the entire revenue of the papacy.[8] The role of alum was so significant within the Holy See that an "alum fund" was created to provide pensions for the clergy.

Nevertheless, even such artful collusion between the church and the Medicis was not enough to contain shadow market forces. Turkish alum continued to enter Europe through smuggling routes. In 1474, an alum glut led to such low prices that Pope Sixtus IV agreed to take a 50 percent cut in royalties. Even so, alum brought incredible wealth to the papacy—enough, according to one estimate, to build five churches on the scale of St. Peter's Basilica. A sign of how important the mordant was to the church's finances is a provision of Pope Leo X's sweeping indulgence of 1517—the one that incensed Martin Luther and ignited the Protestant Reformation. The text offered absolution for "any sin or crime or excess, however grave," with only a few exceptions. In general, these exceptions were for violent crimes against church officials, such as murdering a bishop. Yet included in these unforgiveable trespasses was any attempt to break the ban on outside alum imports!

The papal monopoly on alum was a major economic concern for proponents of the Reformation. As with any bottlenecks in supply, the austerity from this one source led entrepreneurs to seek other sources of alum. Alum deposits were also discovered along the Rio Tinto (Red River) in southern Spain, a region that had been mined for other metal ores for at least three thousand years. Alpine miners had also known that the shale walls of former pyrite mines accumulated small amounts of alum that could be processed like alunite, but the yield was too low to be commercially viable. Then in the sixteenth century, by a comical twist of luck, men relieving themselves against the mine walls discovered that when urine was added to heated rocks, the shale produced ammonium alum.[9] This momentous chemical innovation from humble origins transformed the fate of European royalty. Shale deposits were found in many regions across Bohemia and the Tyrol, which with minor urinary intervention could allow for alum to be crystallized. Coincident with the Reformation, this discovery allowed for the economic liberation of an important commodity for those who dissented with the papacy.

When Henry VIII severed ties with the Catholic Church over his matrimonial misadventures, alum prospecting was a priority for him in the British Isles. The pope had cut off all alum trade to the Tudor court, and the British were left with a colorless repertoire of clothing for much of the latter half of the sixteenth century. Despite numerous efforts, no alum deposits were found within Britain during Henry VIII's lifetime. It would be a few decades after his death that an enterprising naturalist named Thomas Chaloner, trekking through the region of Guisborough in North Yorkshire, happened upon some shale outcrops. He noticed that they were reminiscent of alum-bearing shale he had encountered in travels across Europe. Chaloner's instincts on the rocks proved correct, and this region of Yorkshire soon became the alum heartland of the isles. As with the European shales, this alum also required urine for processing, and the British became masters at procuring an efficient economy of human waste for this purpose. At the peak of alum production in the eighteenth century, the industry required around two hundred tons of urine every year, equivalent to the annual output of one thousand people. The demand for the putrid liquor was such that it was imported from London and Newcastle. Buckets were left on street corners for collection, and reportedly public toilets were built in Hull to supply the alum works. Mixtures of urine, seaweed, and the shale ore were left in the open until alum crystals settled out, ready to be removed. A bizarre method was employed to judge when the optimum amount of alum had been extracted from the liquor—when it was ready, an egg could be floated in the solution.[10] The industry was a remarkable convergence of biotic and geological processes. Alum thus gave us the curious profession of "urine trading" until the late nineteenth century, when synthetic dyes began to be manufactured. The lives of these Yorkshire moor communities were tough and turbulent and even inspired a recent novel by Stephen Chance, *The Alum Maker's Secret*.[11]

Alum's real secret lay in its elemental composition, which remained hidden for much of its heyday as a precious papal resource and beyond. The church had not encouraged scientific research into its properties, nor had the aristocrats who championed the marvels of its bonding to colors supported any investigation of its composition. They were content just to see alum's end products, no matter how putrid and intriguing the interventions might be to get to those results. Nevertheless, the relative abundance of alum during the seventeenth and eighteenth century led to greater interest among scientists in its underlying chemistry. Laboratories in the eighteenth century were eager to take alum apart to see if the miraculous substance had any hidden metallic secrets. Discovering the composition of alum was also of interest to pharmacists, who had been using it in a variety of therapeutics.

The rock seemed blessed to multitudes of Catholics, given its connections to the Catholic Church's fortunes during the Renaissance. Yet alum and, in turn, aluminum's direct effects on the human body were to remain enigmatic and contested for many more centuries. With its wide range of alkaline compounds and salts in the earth's crust, could aluminum have been a building block in the evolution of life, or did the emergence of life play a role in its current mineral distribution?

2

SOIL WITHOUT SOUL

Why Aluminum Was Rejected by Life

What we observe as material bodies and forces are nothing but shapes and variations in the structure of space.
—Erwin Schrodinger, *What Is Life?*

O f all the elements, the one most directly associated with life on this planet is undoubtedly carbon. "Organisms"—fundamental units of life—are derived from "organic" compounds—which in turn connote the chemistry of carbon (periodic table code C). The geologist Robert Hazen lyrically writes about carbon's place in the universe and suggests that the creative orchestra of the cosmos perennially plays a "*Symphony in C.*" The role of carbon in forming complex molecules is singularly significant in crafting the building blocks of life. It is thus not surprising that when scientists began to think through various plausible scenarios for the evolution of life, carbon chemistry was central. One of the founders of modern microbiology, Louis Pasteur, convincingly dispensed with the notion of spontaneous generation of life from nonlife via an "essence" or "soul." He did so with a simple experiment in 1869 by showing that sterile grape juice could never convert to wine on its own. There needed to be some injection of living organisms—like yeast—to start the fermentation process.

Although Pasteur astutely saved science from the vagaries of spontaneous generation, he also paved the path for a somewhat simplistic carbon determinism that persists in many circles today. The insights from the viticulture experiments of Pasteur led us toward what came to be known as "vitalism"—which argued that carbon-based molecules such as amino acids had the specific, innate property to develop life. While true at the level of the proximate synthesis of key life-giving molecules like DNA, vitalism was misleading in terms of the ultimate origin story of life. Vitalism provided an explanation to some degree for how chemistry led to biology—but not for how geology led to chemistry. Furthermore, to understand how geology and chemistry conspired to create life, we would need to bring physics to the party, too. As Erwin Schrodinger noted in 1944 in his seminal treatise *What Is Life?*—all living beings share a common physical feature—they self-organize into structures from highly distributed chemical building blocks. In the language of physics, the development of life decreases entropy.

When the vitalists' vision was adopted by biochemists, there was a general view that a primordial soup of simple carbon compounds like methane or carbon dioxide could combine with ammonia under the effect of radiation to form amino acids. These amino acids constitute the basis of the genetic material needed for the chemical evolution of living forms. The Russian chemist Alexander Operlin (1894–1980) wrote a book titled *The Origins of Life* in 1922 wherein he postulated such a trajectory for carbon-based life's development. Although he did not test these experiments, his work was embraced by the Soviet regime, because it metaphorically echoed the views of Marx's "dialectical materialism," wherein useful material could emerge from a communal exchange of resources in a polity. On the other side of the Iron Curtain, the British chemist J. B. S Hildane also supported Operlin's views of the origins of life, as well as his Marxist pedigree. Hildane, who had once famously brought a jar of urine to dinner at Cambridge to provoke a scientific conversation, was not welcome in the United Kingdom for his unconventional views and demeanor. He immigrated to the subcontinent and died there as an Indian citizen in 1969, where his body was donated to the Rangaraya Medical College for research.

Operlin and Hildane's insights were noticed in the United States, despite the Cold War. Stanley Miller and Harold Urey, researchers at the University of Chicago, began a series of experiments to test if the primordial soup could be created in a laboratory setting. Water, methane, ammonia, and hydrogen were added to a flask, and with the application of charge, amino acids could be formed. However, this experiment's ingenuity was still confined to a closed system, and the second law of thermodynamics suggests that the lowering of entropy in such systems is highly improbable and becomes more so as the number of components increases. Early Earth was by no means a closed system in the context of such chemosynthetic possibilities. While such reactions may possibly explain how organic molecules could arise within the protected membrane of a cell, the initial prospect for the rise of complex molecules in an open system became less convincing. Here was an example of "path dependence" in scientific research, where focusing on the primordial-soup hypothesis led to an empirical outcome that was internally consistent but still not applicable to the broader planetary system.

Connecting physics to geology suggested that the only way life could have evolved was under environmental conditions that allowed for low-entropy states to emerge in an overall context that would still increase the full entropy of a system. Only then could the second law of thermodynamics be maintained. The primordial-soup process did not provide that ultimate condition, although some proximate conditions in specific organic synthesis may have occurred. Weather systems like tornadoes and hurricanes are examples of such an emergence of lower-entropy subregions through stochastic processes. What conditions mimicked such disequilibria, and what elements could facilitate the emergence of life-giving molecules? The organic determinists were missing out on the role of metals in the origin story of existence.

METALLIC MEDIATION

The electronic structure of carbon is responsible for its unique ability of molecular profusion through multiple bonds and chains. Metals were thought to be mere geological spectators in the emergence of life until their own electronic structure revealed characteristics crucial for lower-entropy forms to emerge. At the atomic level, metals have a proclivity to shed electrons, producing positively charged metallic ions. Some metals also have multiple levels of electron loss, offering a range of stable ionic options. It is this ability of metals to shed electrons that gives them properties like conductivity and can create conditions whereby they can catalyze reactions. As the Nobel laureate Ilya Prigogine showed, complexity arises from chemical disequilibrium, and life is the ultimate outcome of the complex agency of molecules. The role of metals in the original emergence of life has become more salient in recent years as we have come to further understand the physical features of the earth that would be most conducive to such emergence. Of particular note are the "transition metals," which form the middle corpus of the periodic table, because of their outer orbitals, in which electrons are in the process of being filled toward a stable state (hence the term "transition").

My training as a chemist during my undergraduate years led me to investigate what the latest research was around the role of minerals in the origin of life. While we know that around eighteen elements were essential for the emergence of life, less is known about why particular elements were *excluded* by chemical evolution. If metals, particularly those with multiple electron-loss potential, or "valence," provide an opportunity for biological agency, why was aluminum rejected by life—despite being the most abundant metal in the earth's crust? By addressing this question, could we also begin to understand how certain elements should be incorporated within human synthetic systems?

This question led me to the work of a laboratory in France that has been challenging the conventional wisdom on the origins of life from a primordial soup—suggesting instead, with increasing levels of empirical and theoretical persuasiveness, that metals mediated our existence. The Evolution of Bioenergetics lab (Bioegnergitique et ingeniere des proteines) at Aix Marseilles University has been working to eliminate the dichotomy between organic and inorganic chemistry in the emergence of life for the past fifteen years. After reading one of their papers published in a relatively obscure anthology, *Metals, Microbes, and Minerals*, I reached out to the lead author, Simon Duval, to ask him why aluminum was absent in the emergence of life despite its abundance and its multiple valences of electron transfer. He began the conversation by noting that "life mainly draws its energy from collapsing electrochemical disequilibria," which are manifest in particular processes of electron loss and gain (oxidation and reduction, or "redox," reactions). Duval and his colleagues have convincingly shown that quinones and flavins are two sets of organic compounds that play a pivotal role as the gatekeepers of the multielectron transfers that some key metals are able to initiate. Transfer of one or two electrons works out well, but three-electron transfers are far more challenging to mediate via these compounds. Aluminum generally undertakes three electron transfers, and in the cases where it can do a two-electron transfer, the compounds formed of the element are not soluble and hence less bioavailable for reactions. Thus, even though Earth's surface offers an abundant source of aluminum, it is not possible to galvanize metabolic processes from the metal.

Duval's work is summarized in figure 2.1, which is derived from a paper published by his team[1] where he notes the key role of "electron gates" played by quinones and flavins. These simple aromatic organic molecules, as well as enzymes, could have formed in the "primordial soup," but they needed energy gradients for agency. Quinones are now also being researched as key materials for metal-free batteries. Once the molecular assembly line is in place, energy transfer can then efficiently occur toward the emergence of life-giving processes such as the "charging up" of compounds like adenosine triphosphate (ATP), the core molecular energy carrier in cells. Metals likely played a pivotal role in hastening the assembly lines that led to life and in adding greater potential for cellular complexity to emerge. One of the ways we are fairly confident about the role of metals is by analyzing the genes held in common between ancient microbes and higher forms of life. These genes give us some indication of the genome of the Last Universal Common Ancestor (LUCA)—an elusive organism that was not the first life form per se but the earliest progenitor whose genes have been retained in its multifarious progeny. The proteins—especially the catalytic enzymes—that have persisted in life since LUCA provide us some clues on the role of metals in the emergence of life. Many

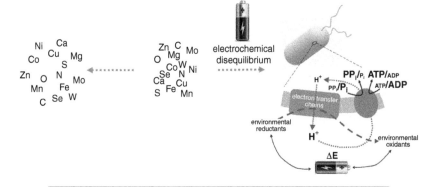

FIGURE 2.1 How a random agglomeration of elements can lead to life through disequilibrium processes. The electrochemical disequilibrium symbolized by the battery requires metallic intervention.

Source: Figure courtesy of S. Duval, Aix Marseilles University, France, 2020.

of these enzymes have metallic centers made of iron, nickel, and molybdenum.

Energy gradients and water-soluble reactions were needed for the emergence of life: the first could either be supplied via sunlight or through chemical energy. In the primordial oceans of our planet, chemical reactions could have occurred at the surface, with sunlight spurring chemical evolution. However, it is now widely understood that during this period of the Hadean epoch there was too much ultraviolet radiation reaching the surface to allow for complex biotic molecules to remain intact. More likely, chemical reactions occurred in the deep ocean—shielded from sunlight—where energy gradients created by the specific temperatures and alkalinity of hydrothermal vents provided the most efficient thermodynamic pathway for the emergence of life. Metals were abundant in this environment as well, providing the electron delivery needed for the reactions to proceed quickly.

To further understand the role of metals in hydrothermal vents, I had a conversation with the geochemist Michael Russell, the lead author for a paper in the *Philosophical Transactions of the Royal Society* titled "The Inevitable Journey to Being."[2] At the time of our conversation in August 2021, we were both in northern Italy—Russell was retired and living in the outskirts of Milan, following a distinguished career as senior research scientist at Caltech and NASA's Jet Propulsion Laboratory. I was in Bellagio for an author's residency at the Rockefeller Foundation's Villa Serbelloni. Much of Russell's research has involved developing elegant chemical models about distant worlds in space and time. NASA's interest in his work stemmed from the need to understand how life could evolve on other planets from a mineral template. Russell relayed to me his frustration that scientific orthodoxy has favored an organic view of life's emergence that has kept metals at the periphery. Even his seminal work on chemical reactions in hydrothermal vents had been appropriated by organic chemists intent on showing how carbon-based citric acid cycles or methane-generated processes could instead lead to the emergence of life.

A crucial dividing line between Russell's work and that of other theorists has been the degree of importance that should be given to geology as the determining factor in biotic evolution. Russell and colleagues in the earth sciences have considered the mineral composition of the early Earth in agonizing detail. Of particular interest to them is the iron-based mineral known as "green rust" and its ability to create energy gradients and attract additional molecules within its physical structure to spur particular reactions. Furthermore, the mineral serpentine, which comprises a porous array of magnesium silicate lattices, could have provided the requisite conditions for the upscaling of the energy gradient, as well as offering the protective enclosures to safeguard life's emergence. Serpentine is California's state rock, and Russell, who spent much of his professional life in Pasadena, suggests that it could be "life's mother engine."

The role of rocks in life's origins seems intuitive at one level, but the dichotomy between "inorganic" minerals and "organic" compounds persists. Chemistries of carbon that revolve around methane and carbon dioxide seem self-sustaining to many biotic chemists who see minerals as mere spectators or late arrivals to the dance floor of life's party. It is thus not surprising that when the journal *Nature* did a profile of Russell in 2011, they referred to him as "Nascence Man," with an artistic rendering depicting an anachronistic polymath clad in Renaissance-era robes and cap. They likened him to "the alchemists of yore . . . taking basic elements and trying to transform them—not into gold, but into the strings of life." Russell takes the dismissals of his theories on the chin and notes that humans have a propensity for "motivated reasoning." He is enthralled by the specter of metals playing a critical role in the emergence of life, but when I asked him about aluminum, he shook his head. Aluminum can impede the emergence of life on green rust—if iron is replaced with aluminum, the electron exchange properties are lost. He recalled memories of childhood in Britain, when aluminum pans were so commonly used for cooking, and said he often wondered about toxicity, given the geochemical rejection of this abundant metal.

Elixirs in Aluminum

Over evolutionary time, living organisms have found a way to metabolize a range of metals and incorporate them into their molecular architecture. Yet despite its massive abundance in the earth's crust, aluminum is not utilized by biotic systems. It is generally not toxic, but in certain forms it can interfere with cellular activity and has also spawned a wide set of activist science around linkages to a range of afflictions including allergies, autoimmune disorders, dementia, and breast cancer. The rejection of aluminum by natural selection in the development of living forms leads us to consider some broader questions about how evolution operates at the elemental level. Which elements are crucial for life, which are toxic, and which are inert in living systems? These are important guiding questions for ecological planning and when choosing which materials to use in particular technologies.

The absence of aluminum in biotic forms is partially explained by its electron valence, which does not favor the kind of catalytic properties that iron and magnesium offer. But an inability to catalyze does not imply toxicity per se. Toxicity of the elements usually occurs if they supplant other useful elements in particular compounds or if they inhibit some key chemical reactions that are essential for metabolic activity. This can happen for elements that may otherwise be essential for life, such as iron or magnesium. Until the 1990s, iron poisoning was the leading cause of death from medicine overdoses in children under six years of age. This was before the advent of widely used tamper-resistant caps on multivitamin bottles. Iron poisoning is caused by the corrosive impact of iron salts on the tissues lining the gastrointestinal tract, such as the stomach and intestines. Fluid retention and blood loss can subsequently occur. The same is true for most elements that may be deemed essential in small quantities. Figure 2.2 shows this relationship for essential elements but also for an element like aluminum, which is not needed for physiological processes and hence after a tolerance threshold drops into negative-impact categories. For therapeutic agents, the curve may be shifted in terms of dose tolerance and positive impacts,

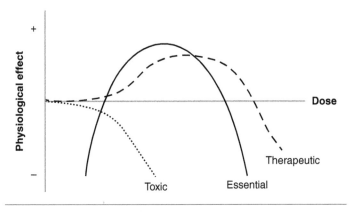

FIGURE 2.2 Variations in elemental toxicity.

but eventually an excess of any material will lead to negative impacts. Even water, which comprises 70 percent of our body mass, is toxic in excessive amounts; soldiers and marathon runners in extreme heat have been known to die of hyperhydration.

Aluminum falls into the category of the downward-sloping "toxic" curve because it lacks any positive physiological value. This lack of positive metabolic value has led to a vast literature that labels it an insidious intrusion in human life. Its geological abundance in the earth's crust has meant that once we developed mechanisms to extract it in economically viable terms, its uses proliferated not only in built infrastructure but also in consumer products. The ancient uses of alum as a mordant had limited exposure in terms of human biological uptake, but its use as an antiperspirant could potentially have exposure pathways via sweat glands. Cookware and utensils of aluminum became fairly common in the early twentieth century, and the ubiquity of aluminum foil continues. Aluminum soda cans containing highly acidic drinks could potentially leach out aluminum, and thus they are coated with a thin plastic. To line the hundred billion beverage cans we Americans swill down every year takes about twenty million gallons of epoxy coating.

The plastic coating on aluminum cans is illustrative of a broader challenge for products where an element is not metabolically useful. Lead, cadmium, mercury, and a number of other elements present similar challenges. However, the economic incentives for using aluminum, given its light weight and durability, are such that industrial chemists have gone to unusual lengths to mitigate corrosive harm. Yet the complexity of human biology has constantly been brought forth to question the safety of these solutions. The range of coating chemicals has changed over the years as health concerns were brought forth about a class of plastics that could disrupt the human endocrine system by mimicking hormones in the body. Of particular concern is a class of compounds called bisphenols, which can mimic estrogen and thereby lead to breast cancer. Various permutations of these compounds, with slight changes made to their molecular structure, have been tried, and most recently Bisphenol F appears to be the safest based on lab animal–exposure studies.[3]

Aluminum has become irreplaceable for the beverage industry. Aluminum cans for carbonated drinks offer ease in transport while maintaining pressure and flavor. Beverage companies have been willing to invest enormous amounts to find alternative coatings rather than find a substitute for aluminum itself. In 2011, the global production capacity of can coatings was estimated to be 800,000 metric tons, which corresponds to a market value of €2.8 billion.[4] The health concerns over coatings led Coca-Cola to enter a confidential research and development pact with six materials firms that collectively spent an estimated $106 million on finding safer coatings. Such was the utility and convenience of aluminum that finding safe coatings to shield it from corrosion—and from bad health-impact publicity—could spur massive investments.

Another aluminum, paper, and plastic container revolution-ized the way we consume food worldwide. In 1943, a young Swedish engineer named Erik Wallenberg got a job at a food packaging company, where he began to experiment with a range of materials to make packaging aseptic. Wallenberg was employed by the industrialist Ruben Rausing, who, inspired by his wife's stuffing of sausages in their kitchen, was developing ideas about airtight packaging. Filing long airtight packaging tubes with liq-uids like a sausage and then truncating the tubes with airtight seals produced tetrahedral cartons that minimized leakage potential and cost. Figure 2.3 shows the layers that led to the development of what is now called "Tetra Pak" packaging, named after the original shape of the distinctive cartons. However, the key to a durable container that could withstand the degradation potential of temperature, light, and oxygen was multiple layers of materials comprising paper, plastics, and, yes, aluminum.

Tetra Pak packaging, combined with ultrahigh temperature (UHT) milk sterilization, transformed nutritional delivery in the developing world. Refrigeration was no longer a necessity for many previously perishable food products. In 2019, the company, now the world's largest food packaging provider, estimated that 68 million children in fifty-six countries have received nutritious beverages like milk in Tetra Pak cartons. And to keep kids fed through the COVID-19 pandemic, Tetra Pak has supported dis-tribution via innovative contactless access to milk through solu-tions such as automated locker systems.[5] Thus the concerns about aluminum's health impact or that of its accompanying plastics need to be considered in the context of such tradeoffs. Using this packaging also reduces the need for other chemical preservatives.

Industrial harmony with human health concerns is thus highly complex, as we continue to learn each time concerns are raised about the risks of particular products. Caution is seductive in

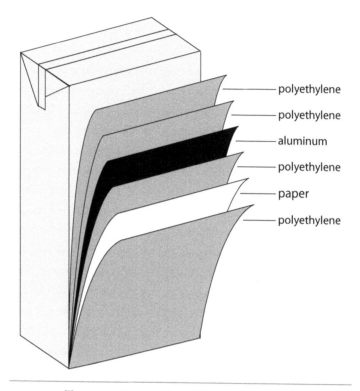

polyethylene

polyethylene

aluminum

polyethylene

paper

polyethylene

FIGURE 2.3 The various layers of Tetra Pak containers. Food contact, top layer, and adhesion layers comprise a range of plastics.

Source: Open-source image from Tetra Pak Corporation.

its appeal—as the aphorism goes, "better safe than sorry"—especially when it comes to health. Yet in a world of complex systems, safety at one point has tradeoffs of safety at another. Activism against particular products is much easier to mobilize when such complexity is not taken into account. At the same time, the industrial status quo is also easy to maintain by pleading complexity as an excuse for inaction. Hard science and even harder social choices confront the academic and the activist in finding liberation from both precautionary paralysis and reck-less chemical misadventures. How health activism around alu-minum products developed provides important insights on the promises and pitfalls of using science for more sustainable decisions.

ELEMENTAL LIBERATORS

"Familiarity breeds contempt"—the proverb is perhaps just as true for the elements we encounter in our daily activities as it is for the people whose paths we cross. As chemicals have become more pronounced in our lives, so too has anxiety increased in many scientific circles about their potential impact on health. Environmentalism often emerges around a feeling of being enslaved by industrial dependence on a particular lifestyle or product. This can be pesticides, or fertilizers, or cars—or indeed aluminum. Elemental anxiety sharpens the resolution of how we may consider those well-intentioned individuals who set out on a mission to liberate us from these shackles of substances. As alu-minum products became more commonplace, concerns about their impact became a rallying cry for activism. But the more diffuse the human contact with an element, the more difficult it becomes to establish any pollution impact causality as well. For a metal like mercury, it was much easier to establish health impact causality because the sources were often easy to track.

When a small fishing village in southern Japan began to show abnormal neurological ailments in the 1950s, mercury was iden-tified fairly quickly as the cause because of a point source of fac-tory pollution that was being bioaccumulated by fish. Mercury salts were being used as catalysts for a range of chemical prod-ucts at the factory, with wastewater discharged directly into the bay, leading to clearly measurable amounts present in fish. With an element like aluminum that is itself so ubiquitous—including its abundant presence in common airborne dust—the causal linkages are far more difficult to establish. Nevertheless, the burgeoning of aluminum products began to raise alarm bells and calls for more definitive research on its possible harms. This was motivated by aluminum's lack of biological functionality. Although it could be rendered inert via some forms of oxidation, there were enough examples of its substituting for iron in biologically functional molecules to merit detailed toxicological studies.

Among the early researchers to conduct detailed studies of aluminum's health impact was a British inventor named James Derek Birchall. The son of a butcher from Lancashire, Birchall left primary school at age fourteen to start work as a laboratory assistant at a local industrial firm. He became an assistant to the chemist Oliver Coligny de Champleur Ellis, who had a knack for making science accessible to the public with books like *A History of Fire and Flame* (1932) and *Poetry and Science* (1924).

The power of great mentors in transforming scientific careers is exemplified by the influence Ellis had on Birchall's life. From humble roots, the young Birchall showed remarkable talent with experimental techniques, and with the support of Ellis and associates, he made his way eventually to the Imperial Chemical Industries (ICI), at the time one of the world's most diversified chemical companies. Birchall worked with a range of salts and dendritic crystals and eventually became interested in fire-extinguishing chemicals. He is credited for developing Monnex—a dry-powder fire extinguisher that was the most effective retardant against gasoline fires and now a staple at most filling stations.

Silica and alumina salts became increasingly of interest to Birchall, and he patented a variety of materials derived from aluminum phosphate, including a plastic film with an aluminum phosphate coating. Birchall was extraordinary in the breadth of his chemical interests, but aluminum became of particular fascination for him. The last twenty years of his life were devoted to understanding the health impacts of dialysis equipment. The puzzle he was trying to solve was why those hospitals in the wealthier suburbs of London, which used sanitized water, were having more health complications than poorer patients in hospitals with ostensibly dirtier water. His research led him to identify the use of aluminum sulfate and related salts of the metal, which were being used as coagulants to sanitize drinking water from contamination by pathogenic bacteria such as *Giardia* and *Cryptosporidium*. Here was an alarming case where the wealthy were being poisoned by their own quest for sterile purity.

So extraordinary were Birchall's achievements in understanding both the positive uses as well as the hazards of the elements that he was given the ultimate British scientific accolade of being elected to the Royal Society in 1981. To this day he is the only twentieth-century scientist without any formal university education to have received this honor. Membership in the Royal Society gave him academic credentials and invitations to lecture at major research universities worldwide. He lamented the distinctions between pure and applied research and saw science as "concerned with the creation of work and wealth through invention and innovation." Courted by various universities, he settled at Keele University in Staffordshire and began to focus his research on aluminum. Before his untimely death in 1997 in a traffic accident, he had befriended a young chemist at Keele University, Christopher Exley, who continued Birchall's quest to liberate the world from the potential toxicity of aluminum, from soda cans to marijuana. Plants have a particularly potent exposure pathway for aluminum, since their roots are in constant contact with soil, which inevitably has a high aluminum content. Plants have evolved defense mechanisms against aluminum toxicity through the expression of the citrate synthase gene in roots and/or the synthesis and release of organic acids that reverse aluminum-induced changes in proteins. However, plants such as cannabis, which are being selectively bred for the bioactive content of compounds such as tetrahydrocannabinol (THC), often do not have such protective measures in adequate amounts. Exley has thus noted concerns about using medical marijuana for treating patients with neurological ailments such as multiple sclerosis. Tobacco plants have also been shown to have a similar uptake of aluminum. Fortunately, it appears that most vegetables we consume as food do not have much metal contamination: thus these protective evolutionary mechanisms have worked well where it matters most in terms of human diet!

Exley, who now refers to himself as "Mr. Aluminum," is convinced that aluminum's chemical activity is incompatible with human metabolism and that it can interfere with other elements such as magnesium, calcium, and iron, which have important metabolic roles. He considers aluminum an "impostor" in our metabolic activity for which we "have not been preprepared through biochemical evolution."[6] Living cells have, however, evolved mechanisms to remove aluminum from within using proteins called "efflux pumps," which have the remarkable ability to selectively bind and expel specific elements. The toxicity of any element is ultimately dependent on the effectiveness of such mechanisms in cleaning up cells. While the ultimate toxicity of aluminum is not questionable, whether the body is able to defend against exposure is where Exley's research gets questioned. Thus, for example, the wide use of aluminum salts in antiperspirants, sunscreens, lipsticks, and other cosmetics could potentially be hazardous, but only if our protective mechanisms on the body are breached.

The mechanism by which aluminum compounds act as antiperspirants involves the release of the metallic ion into the sweat glands, which inhibits the release of perspiration. Exley has campaigned against such cosmetic uses and what he believes is a misrepresentation of how the products actually function. The antiperspirant mechanism is not caused by a physical blockage of sweat pores but by a chemical process whereby aluminum ions can enter the body. Since there are numerous lymph nodes in the armpit, it is conceivable that the ions could then travel to other parts of the lymphatic system and affect other glands in the body. Considerable concerns around linkages between the use of antiperspirants and breast cancer have emerged from this line of thinking. However, the challenge of proving causal mechanisms in complex pathologies with any chemical agent often remains insurmountable.

Animal studies versus human studies are contested given different metabolic rates, and specific cancer clusters are hard to prove because of the many multiple causal factors at the genetic and epigenetic level. Postfactor accumulation of aluminum in tissue of diseased individuals is also difficult to link to causal mechanisms because usually such deposits occur in very elderly patients. For example, aluminum exposure and Alzheimer's disease has been the focus of intense research, and accumulation of the metal in the brain of patients is now widely acknowledged.[7] However, there is a natural accumulation of such plaques in older individuals in any case, so linking aluminum to the disease becomes more difficult to establish. One of the researchers who has challenged Exley's research is Nicholas Priest, who says that aluminum can cause dementia but that such dementia is bio-chemically reversible, unlike Alzheimer's disease. Both researchers spar through journal articles and letters to the editors and accuse each other of positional biases. In an interview via Zoom, Priest noted that the most important investigative work he has done involves tracing the pathways of aluminum through the body using a radioactive isotope, Al 26, of the element. It is worth noting tangentially that the great physicist Enrico Fermi had done some of the earliest experiments on artificially induced radioactivity on aluminum metal, and this was noted in his Nobel lecture.[8] Through the use of such artificially induced aluminum isotope tracing, Priest could tell exactly where the intake of the aluminum was going. Most aluminum that enters the body is excreted in urine.[9]

The kidney's power as a biological filter to remove toxic aluminum has become the main point of contention between Exley and Priest, even though both agree that urine is a good indicator of lifetime human exposure to aluminum. Certain proteins, such as albumin and transferrin, can bind aluminum, and this can reintroduce it into the body during passage through the kidneys. This can also be the pathway for aluminum entering the body of dialysis patients; using distilled water instead of tap water can prevent this contamination. Ultimately, the debate on aluminum toxicity can be zoomed out to the level of global disease patterns and trends. Can we say that Alzheimer's is becoming more prevalent because of greater aluminum use, or is it that the rapid rise in life expectancy thanks to improved health care has led to a greater chance for the prevalence of such diseases of aging? The debate will continue, because definitive answers in such multiple causality mechanisms are often impossible to achieve. However, the take-home lesson for considering broader questions of industrial harmony with nature pertains to what alternatives are available for applying the precautionary principle. For example, if we are to relinquish aluminum in cooking vessels, we have a range of alternatives that do not have particularly higher economic and ecological costs. However, getting rid of aluminum in Tetra Pak food containers could have a major effect on the delivery of food in stressed environments. Lifecycle analysis (LCA) is a technique that can be applied to calculate in detail such comparative metrics, and this can assist with decisions where the marginal health impacts of the usage of a material can be compared across the full scale and scope of the supply chain.[10]

A similar precautionary conundrum pertains to the use of aluminum adjuvants in vaccines to increase their potency. Exley's work hypothesizes that aluminum's presence elicits an antigen response that can lead to a cascading autoimmune reaction that in turn can lead to autism. However, the chance of this happening, given the vast usage of such adjuvants, is relatively small and has to be weighed against the protective effects of the vaccine on more common and lethal diseases for children. Aluminum adjuvants have been used in vaccines since the 1930s, and the U.S. Centers for Disease Control maintains that we now have enough data to conclude its safety, with limited alternatives available.[11] On the other hand, mercury usage in vaccines, as a sterilizing agent called thimerosal, has been largely discontinued, since alternatives are available and new forms of packaging can themselves prevent contamination. At each step of the way in our relationship with the elements used by industry, we need to pause and go through painstaking choices. The devil is proverbially in the details. Sensationalism and blanket fear of chemicals serve little practical purpose. Credible caution must be coupled with an understanding of the delicate relationships the myriad of molecules on our planet form with one another and in turn with our bodies.

THE CHEMISTRY OF COMPANIONS:
SILICON AND ALUMINUM

As with most pollutants, aluminum salts eventually make their way to waterways, and the organisms in rivers and oceans become their ultimate destination. Metals in particular are said to bioaccumulate in fish. It is no wonder, then, that much of the concern about metal contamination in the human diet pertains to fish. The mercury poisoning of the fishing village of Minamata during the 1950s and 1960s from a factory that used mercury sulfate as a catalyst is a landmark case in point. While metallic mercury at room temperature is not absorbed by the body, its vapor and an organic derivative called methyl mercury are. The residents of Minamata were consuming fish in which methyl mercury had bioaccumulated, and this eventually led to debilitating neurological ailments. This case of mass poisoning through fish consumption led to massive research efforts on a variety of metals that could enter human metabolisms through a pescatarian diet. Expectant mothers were warned all over the world to avoid swordfish and tuna, which have a proclivity for metal bioaccumulation.

In this midst, the Birchall lab at Keele University was researching the bioaccumulation of aluminum in fish. They were noticing that fish were not harmed by the aluminum accumulation, nor was there transference of toxicity to humans. One of the early insights from Birchall's own work suggested that fish were protected from aluminum toxicity by the presence of silicic acid—this was a simple molecule with a silicon atom surrounded by four sets of oxygen and hydrogen atoms. It was found abundantly in natural waters, from the weathering and erosion processes of silicate rocks. The key element in these rocks was, of course, silicon. Exley recounted in his book written during the COVID-19 pandemic and titled *Conversations with Mr. Aluminum* that his only paper in the prestigious journal *Nature* was on this key insight about the protective impact of silicon on aluminum toxicity in fish.[12] There is thus a geochemical balance that the close companionship between silicon and aluminum plays in ecosystems. Aluminosilicate rocks are ubiquitous, and the coexistence of these two elements in strongly bonded materials shows that there is an important chemical affinity between them.

Unlike aluminum, however, silicon does have some metabolic role—albeit the details are not yet well known. There is emerging evidence of silicon's role in the development and structure of both connective and bone tissue. Silicon plays some role in the integrity of nails, hair, and skin as well as in the synthesis of collagen. The element is also linked to orthopedic health and bone resilience. However, perhaps the most intriguing role of silicon, which was suggested by Exley, pertained to a reduction of aluminum accumulation in Alzheimer's disease, as well as immune system health and a reduction of the risk for cardiovascular disease. The presence of silicon has been confirmed in a variety of tissues.[13]

The environmental *availability* of aluminum is thus dependent on silicon's presence in acidic form. This remarkable affinity of aluminum and silicon is demonstrated by the similar structures of their oxides as well as by the synthesis of a remarkable array of natural and synthetic minerals called zeolites, which will be further discussed in chapter 6. This affinity between silicon and aluminum exists also in very dilute solution. These species mediate the bioavailability and cellular toxicity of aluminum, and one of the most interesting arrays of experiments carried out by Exley in this context has been with the detoxifying effects of silica-rich mineral waters. He tested a variety of mineral waters and found the French brand Volvic to be particularly rich in silicon (30mg/L), which is operationally 14mg/L of silicic acid. He then tested the effect of the mineral water on his own aluminum discharge in urine, followed by a clinical trial of Alzheimer's disease patients. The results showed a definitive increase in aluminum discharge after mineral water usage; the results were published in the *Journal of Alzheimer's Disease*. Although he had earlier been ignored by Volvic, the article's publication and press coverage led to a call to Exley from Volvic's parent company, Danone. A contract for research was signed for around $50,000, which would support a more detailed study. Exley claims that six months into the contract, the company decided to withdraw its support without explanation. His view is that the range of aluminum-containing products in other branches of the company would have complicated the situation for Danone.

Exley remained convinced of the curative powers of silicic acid. He found another high-silicon mineral water in Malaysia called Spritzer and procured thousands of liters of its product. The trial involved fifteen Alzheimer's patients and their spouses drinking 1.5 liters of the high-silica mineral water per day for twelve weeks. The aluminum burden was reduced in all fifteen patients, and the cognitive function in 20 percent of the patients improved noticeably. Larger and more scientifically convincing trials have not yet been possible. Silicic acid "detox" remedies have become rife since this work was popularized. However, Exley does not like the term "detox" for such elemental purging, because he "considers silicon-rich mineral waters as a philosophy for living in the Aluminum Age as opposed to classical 'detox' therapy." He drinks 1.5 liters of this water daily and continues to work on doing more direct studies on how this habit could prevent neural degeneration.

Within the human body, most scientists delineate three broad categories of elemental activity: (1) major dietary elements such as calcium, phosphorus, potassium, sulfur, sodium, chlorine, and magnesium; (2) minor dietary or trace elements such as iron, cobalt, copper, zinc, manganese, molybdenum, iodine, bromine, and selenium; and (3) others, considered "ultratrace" elements (from their very low concentration and unclear role)—boron and chromium. Silicon is noticeably absent from these categories, but this may well be changing as more evidence for its bioactivity emerges. Furthermore, the element's similarity to carbon has led to greater research on its availability for being the basis for complex life-giving molecules.

Silicon is near carbon in the periodic table and also has four spare electrons in its outer shell. Thus, each silicon atom could theoretically form the same kind of long-chained bonds that characterize carbon-based organic molecules. Is it possible that aluminum mitigated against such a process for silicon, given their timeless tryst, or could they eventually form biotic molecules on their own? The great futurist fiction writer H. G. Wells—whose ideas in many cases were remarkably prescient—suggested in an 1894 article published in the *Saturday Review* that aluminum and silicate materials could conceivably constitute life: "One is startled towards fantastic imaginings by such a suggestion: visions of silicon-aluminum organisms—why not silicon-aluminum men at once?—wandering through an atmosphere of gaseous sulfur, let us say, by the shores of a sea of liquid iron some thousand degrees or so above the temperature of a blast furnace."[14]

While Wells was perspicacious in predicting a range of technologies such as telephones, lasers, and email, his prediction of an aluminum-and-silicon-based life form is chemically unlikely to materialize. However, it may be possible that silicon could be incorporated within organisms to some degree. In 2016, the research team of the Nobel laureate Frances Arnold, who pioneered techniques of molecular evolution, conducted a game-changing experiment with the thermophilic bacterium *Rhodothermus marinus*, found in hot springs in Iceland. Arnold and her colleagues found that the heme protein, with some interventions, could also generate low levels of organo-silicon compounds. However, this happens rarely and requires the carbon in the carbohydrate metabolism of the bacterium, so it couldn't be viewed as providing any evidence for silicon-only-based life.

In 1891, the German astrophysicist Julius Scheiner first published speculation that silicon could be the basis for life on other planets. This prospect was further elaborated by the chemist James Emerson Reynolds (1844–1920), who gave a speech to the British Association for the Advancement of Science in 1893 (subsequently published in *Nature*) and suggested that the thermal stability of silicon compounds could be suitable for "thermophile" organisms to be resilient.[15] Such prospects are, however, quite dim for a variety of reasons worth considering here in the context of broader lessons for the geochemistry involved in the emergence of life. When carbon is oxidized by organisms during the respiratory process of a terrestrial organism, it becomes carbon dioxide—the infamous greenhouse gas and soda-pop fizzer—which is essentially a waste material that is easy for a creature to remove from its body (even a burp will do it). The oxidation of silicon, however, yields a solid: immediately upon formation, silicon dioxide organizes itself into a lattice in which each silicon atom is surrounded by four oxygen atoms. Exhaling such a substance would pose a major respiratory challenge.

There is a remarkable book called *The Encyclopedia of Extra-terrestrial Life*, in which the authors lay out some of the key parameters for such molecular evolution into life. Organisms must be able to collect, store, and utilize energy from their environment. In carbon-based biota, the basic energy-storage compounds are carbohydrates—chained molecules in which the carbon atoms are linked together and combined with a set proportion of hydrogen and oxygen. A carbohydrate is oxidized to release energy (and the waste products water and carbon dioxide) in a series of controlled steps using molecular catalysts—enzymes that are often large complex protein molecules that can physically accelerate bonding and reactions.

A feature of carbon chemistry that allows for such selective molecular manipulation is what we call "chirality"—this is a property of asymmetry whereby a mirror image of a molecule is distinct and not superimposable on its reflected twin. Silicon's failure to give rise to many compounds that display chirality makes it a less probable element to develop life forms.[16] Even so, science-fiction writers have pondered over such prospects. The original Star Trek series had an episode titled "The Devil in the Dark," in which the planet Janus VI was home to a bizarre silicon-based life form called Horta that tunneled through rock like humans walked through air. There is also the Marvel character Sandman, whose origin myth has him fusing with radioactive sand to gain the shapeshifting properties of molten glass. Perhaps the most plausible of the silicon-based life-form stories is Stanley Weinbaum's 1936 *A Martian Odyssey*, in which a silicon-based creature has a very slow metabolism and is half a million years old. It moves once every ten minutes to deposit a brick—which is its waste material. Weinbaum astutely posited a response to one of the bottlenecks for siliceous life by providing a means for metabolizing energy sources and excreting solid waste material.

The abundance of aluminum and silicon has been an alluring source of conjecture and conviction among scientists and nonscientists alike. Any material so ubiquitous in our lives will make us inquisitive about its benefits and potential hazards. The ways that aluminum intersects with biology is a remarkable tale of evolutionary triumph as well as tacit turmoil. Has humanity managed to make the element far more bioavailable than what our anatomy evolved to endure? Are there antidotes, say, silicon, to this potential toxicological exposure? The constructive role of the chemist in such contexts is to determine which compounds of an element are most useful and least harmful. The possibilities for both positive and negative feedbacks in the complex biology of life may seem endless. However, there are some parameters under which chemical reactions and biophysics operate, as was illustrated by the limitations of any prospect for silicon-based life. Understanding these parameters at the scale of chemical bonds and how they are made or broken is the next point of understanding on our journey toward industrial harmony with nature.

II

PRECIOUS FORCES

Soul of the World! Inspir'd by thee,
The jarring Seeds of Matter did agree,
Thou didst the scatter'd Atoms bind,
Which, by thy Laws of true proportion join'd,
Made up of various Parts one perfect Harmony.

—Nicholas Brady, *Ode to St. Cecilia* (1691; inspired by Robert Boyle's
The Skeptical Chymist)

3

UNBREAKABLE BONDS

The Challenge of Extraction

The chemists are a strange class of mortals, impelled by an almost insane impulse to seek their pleasures amid smoke and vapor, soot and flame, poisons and poverty; yet among all these evils I seem to live so sweetly that may I die if I were to change places with the Persian king.

— Johann Joachim Becher, *Physica subterranea* (1667)

L ightning bolts have been potent symbols of natural power for human civilizations. They are the most visible manifestation of the electromagnetic force, which the ancient Egyptians recognized in eels and manta rays. Static charge was palpable to humans, and its perception is as old as the sense of touch. Abrasive action on amber—"elektron" in ancient Greek—was well known and gave us the root word for electricity. Fiddling around with animal fur or even our own hair could release electrons, giving us a tingle, and from early on we suspected this mysterious force as being somehow fundamental to our existence. However, it was not until the eighteenth century that scientists began to consider the chemical aspects of electricity. Two inventions played a key role in allowing for elemental electricity to be appreciated: the "Large Electrostatic Generator" designed by the Dutch engineer Martin van Marum in 1784 and the zinc-copper pile battery built by Alessandro Volta in 1799. Both these inventors became fascinated with the role of metals and electricity and paved the way for our understanding of chemical bonds.

When scientists recognized that electric charge could reduce metals from their salts—as van Marum inadvertently did with zinc, tin, and antimony—they also began to appreciate the bonding potential of the same forces. The dichotomous virtues of positive and negative polarity became more apparent. The attraction and repulsion of particles and electrodes were the mechanisms by which the wonders of electricity revealed themselves to humanity. Salts, which seemed like irreducible and simple substances, revealed their hidden elemental origins when electric charge was passed through their solutions. Thus began the search for metals that hitherto had not been seen in natural form because of their affinity for other earthly elements. Aluminum was perhaps the most inscrutable of them all: it disguised its abundance so well in multitudes of minerals that took varied forms holding little resemblance with their constituent metal. There were not even specific colors that could be associated with its form, as was the case with iron, nickel, or copper salts. These latter metals had crystal structures that let certain wavelengths of light pass through, to leave impressions on the human retina. We thus perceived that copper salts were mostly blue, nickel salts green, and iron, depending on its valence state, either orange or light green. Aluminum salts did not display such structural similitude. They were ubiquitous but offered no visible fixed character.

The British chemist Sir Humphrey Davy had been fascinated by alumina and its unique properties, which had been found useful and exploited since antiquity. Davy was familiar with the work of the German chemist Andreas Marggraff, who had shown definitively in 1754 that the oxides of calcium and aluminum were distinct in their metallic constituents. This discovery spurred him to seek further ways of isolating the two metals that formed these oxides. Despite his humble origins as the son of a woodcarver in Cornwall, Davy outshone many of his elite contemporaries with his brilliant and innovative use of electrochemistry. In 1807, he successfully electrolyzed alumina with alkaline batteries, but the residue had so much potassium and sodium in the mixture that the metallic aluminum was not distinguishable. The metal he sought was too gregarious to be separated from its companion elements. He kept trying to isolate what he now fervently believed to be an undiscovered element, but each time some new alloyed metal occluded its emergence. He was able to isolate many other metals, such as calcium, barium, magnesium, and strontium, but aluminum eluded him—though he did give it its name. Davy suggested the hidden metal be named *alumium* in 1808 and *aluminium* in his monumental book *Elements of Chemical Philosophy*. The different spelling of the metal on either side of the Atlantic can trace its origins perhaps to Davy's original ambivalence on its naming. The ambition of Davy's magnum opus was ahead of its time, and although by modern standards it had some missteps, like its section on "radiant or ethereal matter," it was the first attempt at harmonizing a theory of chemical reactions. Its long enumeration of elements—some of which had not even been isolated, such as aluminum—was a prescient prelude to the organization of the periodic table half a century later.

Davy devoted over 1,500 words to aluminum in his manuscript and described his agonizing attempts to isolate aluminum metal. His astute understanding of chemical bonds and the combustion behavior of alumina also led him to discuss the role of oxygen (or oygene, as it was then called) in alumina's composition. Also noted in his description was aluminum's presence in ruby crystals of corundum. The color, however, was not coming from the aluminum atoms but rather from other impurities, such as chromium. Indeed, blue sapphires are also chemically aluminum oxide corundum but with titanium impurities. The aluminum-oxygen bond is strong, and its tenacity made it hard to extricate the metal. Davy died without achieving his objective, but his mission was carried on by Hans Christian Oersted in Denmark and Friedrich Wöhler in Germany. Both claimed to have isolated the metal, but without clear analytical tools it was almost impossible to determine if the pure metal had been isolated or if an assemblage of metals bonded together—an alloy—was the resultant product.

Metals that comprised alloys were not chemically bonded in the same way as elements in a compound such as aluminum oxide. Instead, alloys were like dissolved salts in a solute—a kind of solid solution or "admixture" with distinct physical properties of strength and conductivity. The most celebrated of alloys, bronze, which had given us an entire civilizational age, was an admixture of tin and copper. It was easier to isolate an alloy like bronze than it was even to isolate a metal like iron—that is why the Iron Age came after the Bronze Age. Elemental bonds defined our material culture in ways that continue to this day as we seek out novel industrial applications.

The isolation of aluminum as a metal became an obsession for Friedrich Wöhler, who was then a professor at the University of Gottingen in Germany. From his initial experiments and repeated trials and errors with alloys, it took him another eighteen years to create small balls of solidified molten aluminum (globules) in 1845. He did so by reacting metallic potassium with aluminum chloride salt. This is where Davy's contributions to the discovery should again be noted, for he had isolated potassium metal using electrolysis in 1807. Potassium, like aluminum, does not occur in metallic form naturally, as it too is highly reactive. Indeed, potassium, lithium, and sodium metals—all in the "alkali" group of metals in the periodic table—when added to water have explosive reactions. Potassium's extreme affinity for chlorine led it to preferentially bond as potassium chloride salt, liberating aluminum metal in the process. The liberation of aluminum by Wöhler also gave him insights about other chemical bonds, many of which were previously thought to be inviolable for mortals to manipulate.

THE ILLUSORY BONDS OF LIFE

The nineteenth century was a time when chemistry was largely concerned with what was deemed "inanimate"—nonliving. Biology was largely divorced from chemistry, as there was a strong belief that the compounds that emanated from life were beyond the capability of human synthesis. "Vitalism"—a belief that the origin and phenomena of living cells are reliant on a force distinct from purely chemical or physical ones—was the dominant view. The molecules emanating from living systems, such as blood or urine, are indeed complex. Around the same time that he was toiling to isolate aluminum metal, Friedrich Wöhler was also interested in the structure of molecules. He had experimented with salts of silver and, alongside a rival chemist, Justis Liebig, had found some surprising and paradoxical results. The salts "silver fulminate" and "silver cyanate" had the same chemical formulae but highly divergent properties—the former is explosive, while the latter is a stable solid at room temperature. Wöhler had been mentored by the great Swedish chemist Jöns

Jacob Berzelius (1779–1849), who had also observed a strange divergence of properties between salts of tin with identical formulas. The only way to reconcile these properties was to consider that elements in a molecule could be assembled in different ways and exhibit different properties merely because of their physical arrangement. This phenomenon was termed "isomerism" by Berzelius. Isomers are forms of a molecule where the chemical formula comprising all constituent elements in a compound is the same but different structures can be created based on where the bonding of each element occurs.

Working with silver cyanate led Wöhler to consider the arrangement of carbon and nitrogen in another compound, ammonium cyanate, whose elemental composition (CH_4N_2O) was the same as urea—a compound excreted in the urine of animals. Wöhler was fascinated by this similarity because urea was considered a biotic molecule, and thus its laboratory synthesis would have challenged the tenets of vitalism—that biological molecules are of an entirely different order and are formed only with the contribution of an ineffable life force. However, he did not set forth to discredit vitalism per se; his primary interest lay in understanding isomerism. He experimented with a variety of metallic cyanate salts, including those of lead and mercury, alongside ammonia and was eventually successful in synthesizing urea. The gravity of his discovery was evident to him when he wrote to Berzelius in 1830: "In a manner of speaking, I can no longer hold my chemical water. I must tell you that I can make urea without the use of kidneys of any animal, be it man or dog."[1]

Nevertheless, the urea made by Wöhler was not constructed via the same pathway that biotic systems utilize. The complicated metabolic process for urea synthesis in mammalian organs was discovered almost a century later by Hans Krebs and his associate Kurt Henseleit. This reaction plays out in the liver or the kidney and involves six different enzymes. In that sense, Wöhler's synthesis was not a definitive discrediting of vitalism—nor did he assert it was—but it certainly made a dent in the theory. The multitudinous ways that chemical bonds can form or can be broken in natural systems is the key take-home message from Wöhler's work with aluminum as well as with urea. The urea synthesis is often described as heralding both the distinct field of "organic chemistry" and the unification of biology and inorganic chemistry. "Organic chemistry" was to take its name from living organisms but was later defined as the chemistry of carbon—the element that constitutes the complex chemical bonding structure of all biotic molecules.

Vitalism was challenged by the Wöhler synthesis but endured, and to this day there are still supporters of the theory. The biophysics of chemical bonds, particularly involving metals, remains a contentious area for research on the origins of life, as was noted in chapter 2. Scholars continue to debate when molecular reductionism and emergence theory finally trumped vitalism. The historian of science Peter Ramberg devoted a ten-thousand-word article to dispelling the "Wöhler Myth" of the demise of vitalism with urea synthesis.[2] There are passionate views on this topic, as it also tends to link with views about the role of the elements in religious doctrines and the tensions between creationism and established science. However, there are some ways that the mechanistic process of bond building and "vitalism" can potentially be reconciled with modern science.

To understand the illusory nature of the boundary between chemistry and biology, we may need the help of that most fundamental of sciences—physics. It is through physics that we can understand how the flow of energy can transform the inanimate into the animate. The field of biophysics has emerged as an important means of understanding how the elements can coalesce and lead to energy flows in living organisms. Among the doyens of biophysics and one of the great communicators of the Caltech classroom—Rob Philips—has dared to use the term "molecular vitalism" in some of his lectures to describe the energetics of living cells. In one of his lectures at the Kavli Institute for Theoretical Physics in 2019, Professor Phillips noted how the massive amount of energy transfer in living forms is a key characteristic of molecular vitalism. Using nothing more than the traditional blackboard and colored chalks, he enthralled the audience by calculating on the fly the amount of power generated per kilogram by a bacterium, a human, and the sun. Astonishingly, the sun—an inanimate, giant, nuclear fusion reactor—produces around 0.0001 watts per kilogram; a human being produces around one watt per kilogram, and a bacterium produces around *one thousand* watts per kilogram. He further went on to illustrate the role that metallic ion gradients play in these processes. Energy transfer within living forms requires a remarkable amount of energy transfer, which comes from multitudinous chemical reactions.

The mere action of moving my finger to type this sentence has required innumerable chemical reactions and energy transfers between the original intention generated in my brain's neurons and the messaging carried forth to my muscles. Figure 3.1 shows a diagram that illustrates why isolating aluminum was such a fundamental challenge as well as why certain elements are more suitable for biochemical reactions and, indeed, why some industries may be more sustainable. "Free energy" refers to the amount of energy not locked up in bonds and available to do work. Natural systems as well as industrial systems have chemical reactions that try to mitigate the energy needed to reach the desired products. The role of catalysts is to reduce this energy required. In biological systems, catalysts are often complex enzymes, which may be used in industry, too. Transition metals

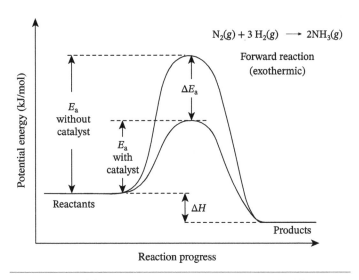

FIGURE 3.1 The reaction progress diagram showing the "Gibbs Free Energy" changes needed to spark chemical reactions (activation energy E_a) with and without a catalyst.

are also effective catalysts in many industrial processes because they can modulate energy flow given their outer electron configurations. Much industrial development has been made possible thanks to the role of catalysts in reducing the activation energy for reactions. At least ten Nobel prizes in chemistry have been given to research related to catalysis in some form. The 2022 chemistry laureate Benjamin List noted in his award lecture that he estimates around one-third of the world's GDP is linked to the use of chemical catalysts.

Energy stored in chemical bonds can be harnessed to conduct useful work by living organisms. Although Einstein famously made us aware that at fundamental physical scales mass and energy have convertibility, for all practical purposes we can say that energy is "conserved" under most conditions on Earth and merely changed from one form to another. Under conditions of time symmetry (meaning that the physical laws of gravity have remained the same between, let's say, the dropping of a ball a thousand years ago versus dropping one today), energy would be conserved. This fundamental insight came to us from the neglected mathematical genius Emmy Noether and her work on linking symmetries in nature to conservation laws in the first quarter of the twentieth century. The functional conservation of energy that would drive our development of industrial processes was noted a few decades earlier by the inquisitive medical doctor Julius Robert von Mayer, who was intrigued by observations he made on patients while voyaging as a maritime doctor in the Dutch East Indies. At the time, bloodletting was part of medical practice, used to relieve the body of perceived pollutants, and Mayer noticed that the blood of patients in the tropics was lighter in color than what he had observed in Germany. Having thought through the need for the body to use food to create warmth, he wondered if people in warmer climates could somehow utilize energy from their environment and "conserve" their bodily energy, thus leading to lighter-colored blood. Although he did not understand the chemical reactions involved, Mayer had inadvertently stumbled upon the importance of the conservation of energy between living and nonliving systems.

Metals also play a role in the color of blood—the iron atom at the core of the massive hemoglobin protein gives it its distinct red color. It is the ability of iron to have a flexible and energetically efficient bonding and debonding mechanism that makes it the choicest metal for this process. All invertebrates use hemoglobin for oxygen transport, except for fish from the family *Channichthyidae* (white-blooded ice fish). The fish can live without hemoglobin because they have low rates of energy usage (metabolic rates) and given the high solubility of oxygen in water at the low temperatures of their habitats (the solubility of a gas tends to increase as temperature decreases). No other metal can play the role of iron in this context. However, *Channichthyidae*'s lack of iron-rich hemoglobin means that they can only carry in their blood 10 percent of the oxygen that other invertebrates do, but given the specifics of their habitat, dissolved oxygen in their white blood fluid suffices for the ice fish. But for almost every other form of multicellular animal life, the importance of oxygen for respiration in living systems that is exemplified in life-giving hemoglobin's role also has broader ramifications for our understanding of the essential industrial reactions that would pave the way for the age of aluminum.

THE CONUNDRUM OF COMBUSTION

We now know that fire on Earth is the result of a chemical reaction fueled by oxygen, just as a chemical reaction fueled by oxygen is the source of fuel in the human body to charge up our biological cells. However, this simple reality, which we now take for granted even in middle-school classrooms, eluded human understanding for much of industrial history. The chemistry of fire was an enigma for chemists and led in the Middle Ages to chemical theories as fanciful as those of vitalism. The late seventeenth century was still an "alchemist's world" where the primary empirical goal of extracting pure precious metals led to far-fetched and speculative ideas. How bonds formed or were broken or why smelting produced certain results was not of much consequence to the alchemist. Steeped in myth and conjecture, alchemy was a protoscience tradition that tried to explain fire through the invocation of the distinct combustion properties of an intrinsic substance that came to be known as "phlogiston," from the ancient Greek word φλογισ, for "catching fire."

This notion can be traced back to 1687, when the German alchemist Johann Becher published his book *Physica subterranea*, where the earliest mentions of phlogiston theory can be found. Fire and air were eliminated from the traditional element model, and Becher replaced them with three forms of earth: *terra lapidea* (stony/rocky earth), *terra fluida* (liquid earth), and *terra pinguis* (oily/fatty earth). In this worldview, *terra pinguis* is the "fire-element," what makes materials inflammable. While burning, this elemental property is released and manifests in the air as flames. Residues such as wood ash are "lighter" (less dense) than the parent material, and this was suggested that *terra pinguis* had "escaped." In a similar vein of misguided logic, heating a metal in air produces a "calx," which is "lighter" than the metal in a similar fashion. In 1703, Georg Stahl, a German professor of medicine and chemistry, published an extended version of Becher's theory and brought the phlogiston theory into mainstream currency for much of the eighteenth century. Stahl's student Jonathan Pott further described properties of this elusive "essence" as follows:[3]

- The form of phlogiston consists of a circular movement around its axis.
- When homogeneous, it cannot be consumed or dissipated in fire.
- The reason it causes expansion in most bodies is unknown but not accidental. It is proportional to the compactness of the texture of the bodies or to the intimacy of their constitution.
- The increase of weight during the process of heating is evident only after a long time and is attributable either to the fact that the particles of the body become more compact, decreasing the volume and hence increasing the density, as in the case of lead, or because little heavy particles of air become lodged in the substance.

- Air attracts the phlogiston of bodies.
- When set in motion, phlogiston is the chief active principle in nature of all inanimate bodies.
- It is the basis of colors.
- It is the principal agent in fermentation.

It seemed as if any explicable property of matter was being piled onto phlogiston, particularly when attributing to it such aspects as color and fermenting prowess! Such an approach to science was lazy and soon began to be challenged by curious and empirically driven minds. Since combustion was such a fundamental observable phenomenon, this theory had ramifications for a broad range of chemistry. There were thus ample opportunities to test and question the presence of phlogiston. Among the savants most interested in the relationship between geology and chemistry was the British tycoon Henry Cavendish, who had inherited two vast fortunes that made him the largest depositor in the Bank of England. Like many aristocrats of the eighteenth century, Cavendish was interested in major elemental discoveries. He would engage in a range of experiments at his personal laboratories at Clapham Common in London. Although he is most well known for his experiment to estimate the density of Earth using lead balls and a torsion balance, Cavendish also conducted a series of experiments on metals with acids. In 1766, Cavendish immersed zinc, iron, and tin in sulfuric acid and hydrochloric acid and studied the gas that was released. He found that it was eleven times lighter than common air and highly inflammable. Cavendish concluded "whereas the metallic substances zinc, iron and tin are dissolved in spirits of salts (HCl) or diluted vitriolic acid then phlogiston flies off without having its nature changed by the acid." The gas Cavendish had discovered and was labeling phlogiston was actually hydrogen.

Although he tried to fit its discovery within the context of the established theory of the time, he noted the curiosity of the high combustibility of the gas, which would potentially contravene this theory (as the second point of James Pott's definitional typology had it, phlogiston in its pure state should not ignite).

The intrepid French scientist Lavoisier was also interested in questioning the phlogiston theory. His approach was to consider the changes in weight after combustion. In his diaries he noted his measurements of an experiment that burned sulfur and phosphorus: "Carefully established by experiments that I regard as decisive, led me to think that what is observed in the combustion of sulfur and phosphorus may well take place in the case of all substances that gain weight by combustion and calcinations and I am persuaded that the increase in weight of metallic calxes is due to the same cause." This weight gain was in fact caused by oxidation and the sequestering of oxygen into the burning elements. This was a decisive blow to the phlogiston theory but also the pathway toward a correct explanation for combustion. In further work he showed that air is a mixture of two gases, "vital air," which is essential to the process of burning (and respiration), and *azote* (from the Greek *azoton*, "lifeless"), which did not support either. *Azote* was subsequently labeled nitrogen in English, although it has kept the earlier name in French and several other European languages. Lavoisier named "vital air" *oxygène* in 1777 from the Greek roots *oxys* ("sharp," from the taste of acids) and *-genēs* ("begetter"), from his errant impression that all acids contained oxygen. In fact, it was the element that Cavendish had unwittingly discovered a decade earlier—hydrogen—that was the essential component of acids.

Combustion's secrets were further uncovered when the ability of metals to ignite with their oxides was discovered in the late nineteenth century. Such is the affinity of oxygen for certain metals, such as aluminum, that when a powdered mixture of the metal and rust (iron oxide) are ignited, a flashy, high-energy reaction occurs that can generate a flame fit for welding. The aluminum in this thermite reaction steals the oxygen from the rust and leaves behind a lump of molten iron. Contrary to popular perception, metals can also burn, and their affinity for oxygen determines the heat produced, since there is no release of gases or liquids as byproducts to temper the energy release. That is why metal powders, including aluminum, have been the mainstay of fireworks and pyrotechnics. Thermite reactions have also been used in welding and have found military applications in grenades and incendiary bombs. From thermodynamic calculations, it is estimated that one pound of aluminum fully reacting with oxygen is equivalent to detonating three pounds of TNT.

Long before the advent of thermite reactions, the heat and spark of elements had beguiled chemists into following errant pathways of scientific inquiry. The ability to extract metal from heating certain kinds of rocks had been known since antiquity and gave us the Bronze and Iron Ages. Metals were believed to have lacked phlogiston when reduced to certain "base" points after their salts or ores had been inflamed. Yet the chemical underpinnings of this phenomenon perplexed the alchemists of the Middle Ages. The phlogiston theory was one attempt to understand such phenomena. Like the Burning Bush of the Old Testament, a mysterious essence was conjectured to be the creative spark of existence. The discovery of oxygen's role in combustion and in metallic rusting by Lavoisier in the late eighteenth century put the phlogiston theory to rest. A neglected chemist who contributed to the demise of the theory but who also questioned the counterorthodoxy was the American chemist Elizabeth Hume. Her 1794 book *An Essay on Combustion with a View to a New Art of Dying and Painting, Wherein the Phlogistic and Antiphlogistic Hypotheses Are Proved Erroneous*, was remarkable in its willingness to question the polarity of arguments. Aluminum's propensity for thermite reactions is an even more damning repudiation of the phlogiston fantasy.

The demise of the phlogiston theory heralded what is now called the "first chemical revolution." Understanding the relationship between combustion processes, metallic corrosion, and oxygen unified the concept of reactions as elemental processes that could take place without any need to posit additional abstract entities like ethers or essences. Such a view also had broad implications for how we would later consider sustainability within the parameters of the law of conservation of mass. Which elements could resist corrosion by oxygen and acids and thereby offer greater industrial opportunities for use, reuse, and recycling? Industry would begin to focus on ways to craft alloys and other compounds that could develop such material resilience. It was in this context that the properties of aluminum began to shine through its elemental competitors.

THE MIRACLE METAL

The advent of the chemical revolution would also give rise to the differentiated salience of elements—particularly metals. Aristotle's idea of the classical elements of earth, air, water, and fire was superseded most notably by the Irish chemist Robert Boyle in his 1661 book *The Sceptical Chymist, or Chymico-Physical Doubts & Paradoxes*. Written as a series of dialogues between five friends with differing philosophical views, this revolutionary work advocated that all matter was composed of "corpuscles"—which in turn could be made up of atoms. In his subsequent work *The*

Origin of Forms and Qualities (1666), Boyle used corpuscularianism to explain what would lead to key foundations of elemental chemistry:

- The idea that compounds can have secondary properties that differ from the properties of the elements that are combined to make them became the basis of molecular identification.
- The idea that the same elements can be predictably combined in different ratios using different methods to create compounds with radically different properties became the basis of stoichiometry and crystallography.
- The ability of chemical processes to alter the composition of an object without significantly altering its form is the basis of mineralization and the understanding of numerous metallurgical, biological, and geological processes.

The discovery of aluminum offers a convergent example of these properties. Boyle was conscious of the unique properties of alum and referred to it as a "perfectly mixt body." This was an astute observation, despite a lack of clear empirical evidence about the bonding strength of aluminum's salts. The metal itself barely exists in free form in nature because of its strong affinity for other elements, such as oxygen. Only a few isolated examples of "native aluminum" existing in flaked alloy form are known, such the discovery of a specimen entombed within a pegmatite rock vein in Bulgaria or in a fumarole within the Kudriavy volcano in far-eastern Russia's Kuril Islands. In China, the Getang deposit in Guizhou province became a particular curiosity when archeologists in the 1950s found small fragments of native aluminum in the third-century tomb of a Jin-dynasty general named Zhou Cho. Consensus emerged that extraction of the metal was not technologically feasible at that level of purity with the technology of the time, so the metal fragments may well have come from modern-day Guizhou.

Molecules that comprised its combination with oxygen, such as alumina, had widely differing properties from the metal itself. Not only do the aluminum atoms form bonds with other elements to form "corpuscules," but they can also further form ordered crystalline structures with other corpuscules as well. Silicon acts as a sister element to aluminum in this regard, and the oxides of both can form intricate molecular lattices that comprise much of Earth's surface minerals. The wide abundance of aluminosilicate minerals in surface rocks on our planet is an indelible testament to the material marvels of aluminum. These minerals can have vastly different crystalline structures and properties. From gemstones to volcanic glasses, aluminum atoms form much of our planetary crust's mineral edifice. Aluminum's particular properties in mineral form are in some ways even more consequential for life on Earth than its metallic properties in isolation. Its specific bonding strengths, particularly in combination with silicon minerals, gave our planet the hard surface of bedrocks on which much of civilizational development could occur. Feldspars, which constitute almost 60 percent of Earth's crust; granites; and basalts all are rich in aluminum. Furthermore, most of the rocks and their constituent minerals that keep our planet's crust firm owe many of their most important properties to the metal's presence in their mineralogy. Aluminum has thus been instrumental in literally creating the stable foundation for life to develop above the viscous, superheated mantle of the planet.

Ever since its isolation as a metal, aluminum began to captivate the industrial establishment. Technologists saw opportunities with this metal's properties for a wide array of products, yet the challenge remained that despite its mineralogical abundance, it was extremely difficult to extract. Much of the story of industrial innovation and chemical synthesis revolves around the economics of scaling up production. However, the scaling phenomenon is a cascade in which every input step in materials and energy must also be considered. The ingenuity of nineteenth-century chemists to work through this scaling process for aluminum is a hallmark of the industrial history of humanity and spanned many steps.

Following the definitive discovery and isolation of aluminum metal, Henri-Etienne Sainte-Claire Deville, a French chemist and technologist, transferred the chemical method of aluminum creation discovered by Wöhler and produced the first industrial-scale quantity of aluminum in the northern French town of Rouen in 1856. Deville's key innovation was to use sodium metal as the reducing agent to extract aluminum from its complex chloride salts. The gradual easing of aluminum's bond with oxygen required multiple steps and industrial-scale inputs to be feasible. However, sodium metal's production was itself a hugely challenging process given its explosive reactivity with water. The American chemist Hamilton Chester discovered a process to extract sodium metal from molten caustic soda using electrolysis. This reduced the cost of sodium production fivefold, but there were still other inputs to worry about. The pungent and toxic chlorine gas and activated carbon at high temperature were also needed to form the double chloride from which aluminum metal would ultimately be reduced by sodium. In those days, chlorine gas was synthesized by reacting manganese dioxide—naturally occurring in the mineral pyrolusite—with hydrochloric acid. An attraction of this reaction was that, with a couple of intermediate steps, the manganese dioxide could be recovered and recycled. Hydrochloric acid was readily available even in those days as a common acid formed via various methods involving chloride salts.

Aluminum's properties captivated the burgeoning technological world of the late nineteenth century. France was replete with its own industrial marvels and discoveries and riding on the success of the Exposition Universelle in 1855, which was attended by over 5 million visitors. Twelve small aluminum ingots were displayed for the first time to the public at this exhibition. The metal was heralded as *l'argent de l'argile* (the "silver from clay"), referring to its physical similarity to silver but its extraction from abundant bauxite clays. The root of the word "bauxite" also had a French pedigree. The geologist Pierre Berthier had discovered this ore in 1821 near the village of Les Baux-de-Provence in the French Alpilles—hence bauxite. The ore is an agglomeration of various minerals containing aluminum oxides in association with water molecules—most notably gibbsite, diaspore, and boehmite. Gibbsite, which has three water molecules associated with the alumina, is called a "trihydrate" and is the easiest for aluminum extraction purposes.

The formation of bauxite takes place primarily through a weathering of the vast amounts of aluminosilicate rocks that pepper Earth's crust. Such weathering processes are often facilitated by tropical, humid climates. However, over geological time it is important to note that currently cold, arid, and temperate climates, such as that of France, may have had a tropical, humid past. Depending on the accompanying rock types, there are primarily two forms of bauxite—lateritic bauxites with dominant iron and silicate content and karst bauxite ores with dominant limestone (carbonate) content. The carbonate bauxites occur predominantly in Europe and the Caribbean; the lateritic bauxites are primarily found in Australia and Africa.

Unlike silver, which corroded and blackened over time, aluminum seemed impervious to any deterioration from exposure to air or water. A thin and almost transparent layer of aluminum hydroxide protects it from further corrosion while maintaining its aesthetic appeal and overall utility. Indeed, among the more important characteristics of aluminum is its instantaneous formation of this self-limiting, resilient film in its liquid and solid states. This coating is thin enough for it not to obstruct the metal's value while thick enough to protect it from corrosion over long periods of time.

The French emperor was suitably impressed with the metal and promised Deville an unlimited subsidy for aluminum research. It is estimated that the entrepreneur received around 36,000 francs from the French government, which was around twenty times the annual income of a French family at that time. Having endured numerous conflicts since his uncle's defeat at Waterloo, Napoleon III's primary interest in aluminum lay in its military uses for weapons, helmets, and armor. Its light weight, coupled with its strength and potential abundance, made it ideal for the nascent modern military-industrial complex. The shiny allure of the metal was also captured by the finer tastes of the French in developing utensils and jewelry. At a point when the metal had still not been displayed to the public, Napoleon is reputed to have held a banquet where the most honored guests were given aluminum utensils—while the others were relegated to eating with gold cutlery! Aluminum's splendors were noted at the Paris exhibition by writers including Jules Verne and Charles Dickens. Commenting on the success of Deville's commercialization prospects for aluminum, Dickens wrote, "Aluminum may probably send tin to the right about face, drive copper saucepans into penal servitude, and blow up German-silver sky-high into nothing." The Russian writer Chernyshevsky wrote in his novel *What to Do* of the metal's salience to socialism: "One day, aluminum will replace wood and may even replace stone. Look, how luxurious it is all, aluminum is everywhere." The prescience of these literary visionaries was capped by the French science-fiction writer Jules Verne, who described a twenty-thousand-pound aluminum space rocket in his novel *From the Earth to the Moon*. The Apollo 8 capsule that orbited the moon a century later weighed approximately that much.

The marvelous malleability of aluminum, which led Verne to choose the metal not only for space exploration but for Captain Nemo's submarine in *Voyages Extraordinaires* as well, continues to inspire artists and architects to this day. The contemporary computational architect Marc Fornes has dedicated many of his most famous architectural embellishments to works with aluminum. Reminiscent of the submarine, the aluminum-shell structure is a "transporting vessel" that is part of the architect's THEVERYMANY series, dedicated "to making real the fantasies otherwise relegated to the digital and we are practiced in fabricating architecture that defies convention. The drama of the interior atmosphere makes you feel as if you have suddenly descended into some other world, despite the thinness of the enclosure that separates you from the familiar terrain beyond."[24] Yet to reach a point where such large artistic forms could utilize aluminum in copious quantities, a key turning point had to occur. The world production of aluminum in 1884 equaled 3.6 metric tons. Around two hundred tons of the metal were produced in thirty-six years (1855–1890) total. A decade later, at the turn of the century, 6,800 tons were being produced in a single year. Such a dramatic rise in production was only possible through another set of monumental innovations that occurred simultaneously and independently on opposite sides of the Atlantic Ocean.

4

THE BOND BREAKERS AND
THEIR BOUNTY

There is nothing harder than to make people use a new metal.
Luxury items and ornaments cannot be the only sphere of its
application. I hope the time will come when aluminum will
serve to satisfy the daily needs.

—Henri-Etienne Sainte-Claire Deville

Industrial enterprise is hinged on scale. Even the most incredible and world-changing discoveries can be of little consequence if they cannot be replicated with efficiency and made available to broader segments of society. The discovery and isolation of aluminum metal exemplifies the importance of scale and its ramifications for sustainability. Although Deville's work and the support of Emperor Napoleon III firmly established aluminum within the pantheon of industrial metals, the economies of scale were still not favorable for its widespread competitive use against iron or copper. However, the interest in the metal remained widespread and spurred creative impulses from many scholarly directions. In 1889, two articles were published on the importance of aluminum's industrial synthesis, one in the *Journal of the Society of Arts* and another in the journal

Science, and both conferred on aluminum the properties of a "noble" and "precious" metal.[1] A discussion regarding the constituent synthesis of the metal by the Aluminum Company of Britain, near Oldbury, was documented in a magazine. For the first time, the inputs and outputs of the industrial process were presented in a public forum, as shown in figure 4.1, a reproduction of the published paper from 1889.

This recipe for aluminum's extraction was written at a time when the metal was still deemed precious. In 1884, the price of aluminum was approximately $1 per ounce, the same as the

```
         Metallic sodium ........      6,300 lbs.
         Double chloride ........     22,400  ,,
         Cryolite  .............      8,000  ,,
         Coal .................         8 tons.

    To produce 6,300 lbs. of sodium is re-
quired:—
         Caustic soda  ..........    44,000  lbs.
         Carbide made from pitch, ⎫
           12,000 lbs. ..........  ⎬  7,000   ,,
         Iron turnings, 1,000 lbs. ⎭
         Crucible castings ......       2½ tons.
         Coal ......  ..........        75   ,,

    For the production of 22,400 lbs. double
chloride is required :—
         Common salt ..........      8,000 lbs.
         Alumina hydrate........    11,000  ,,
         Chlorine gas ..........    15,000  ,,
         Coal .................       180 tons.

    For the production of 15,000 lbs. of chlorine
gas is required:—
         Hydrochloric acid ......  180,000 lbs.
         Limestone dust ........    15,000  ,,
         Lime ................      30,000  ,,
         Loss of manganese ....     1,000  ,,
```

FIGURE 4.1   Ingredients needed to produce one ton of aluminum in 1889.

Source: From extract published in the *Journal of the Arts*, March 15, 1889.

market price of silver. The world production of mined silver in 1884 was approximately 2,834 tons. Best reported estimates for world aluminum production in 1884 total 3.6 tons, primarily in France and England, with some in Germany. In the United States, the only aluminum works belonged to William Frishmuth, who had been a close associate of Abraham Lincoln. He established a foundry in Philadelphia that produced fifty-one kilograms in 1884. Frishmuth was a German immigrant who had also been a student of Frederick Wöhler at the University of Gottingen and shared his passion for aluminum. The Deville process was expensive, given the use of sodium, and limited the overall production of the metal, but Frishmuth was determined to find a way to elevate aluminum in the pantheon of metals. By one account, he spent over $53,000 of his own money (far more in today's equivalent) over twenty-eight years to purify aluminum for casting.[2]

Aluminum was considered a relatively precious metal across the Atlantic as well, but that is not necessarily why it was chosen to adorn one of the most famous honors in the annals of American history. In 1848, after several decades of deliberations on design and scale, Congress agreed to commemorate the first president of the nation with a monument. Rather than installing a bronze statue of the president on a horse, the typical tribute of those times, the decision was made to erect the world's tallest obelisk in the middle of the sprawling National Mall that epitomized the grand urbanism of Washington's great city planner Frederick Law Olmstead. A further thirty-seven years passed before the monument was officially opened in 1885. More than a century later, the metallurgist George Binczewski carefully documented the history of why aluminum had been selected for the obelisk's tip by examining records in the National Archives. He determined that the choice was not because of the perceived

value of the metal. Copper, bronze, or brass, plated with platinum, were the materials preferred by the Army Corps of Engineers, whose chief engineer, Colonel Thomas Lincoln Casey,
oversaw the construction.

Aluminum was chosen as partly because of its weight and
carriage—and because of the influence of William Frishmuth,
who appears to have held considerable political clout in Washington.[3] He wanted to use the monument to popularize aluminum and his products. He assured Casey that he would be able
to complete the construction economically; if not, he would
revert to copper alloys. South Carolina corundum was used as
the aluminum ore to add a symbolic sense of post–Civil War
camaraderie with the erstwhile Confederacy. Even though the
eventually negotiated cost of $230 ended up being almost three
times what Frishmuth had initially quoted, Casey reluctantly
accepted the final product. When the aluminum pyramid was
completed with the setting of the top stone in December 1884,
there was massive publicity on the choice of the metal. The purpose of the cap was to act as a lightning conductor, and initially
it seemed that aluminum would do this task just fine. However,
a thunderstorm in June 1885 damaged part of the cap and its
adjoining masonry, which led to some copper rods being added
to the cap. The additional copper appendages were studded every
1.5 meters of their lengths with copper points more than 7 cen-
timeters in length, gold plated, and tipped with platinum.

Around the same time that this modified version of the cap
was being attached, another major metallic monument was crossing
the Atlantic. The French wanted to celebrate America's emancipation of slaves with an unprecedented gift—a massive copper
statue of the Roman goddess Libertas. The staggering monument was sculpted by the great artist Frédéric Auguste Bar-
tholdi upon a metal scaffolding designed by Gustave Eiffel,

whose eponymous tower is still synonymous with Paris. The Statue of Liberty was dedicated in New York on October 28, 1886. But even as the political bonds of friendship were being strengthened between France and the United States, an intense metallurgic competition would arise the same year. This rivalry would break the chemical bonds and liberate that most abundant metal in our crust in ways that would truly free aluminum to become ubiquitous in our products.

A CONFLUENCE OF
TRANSATLANTIC DISCOVERY

Upscaling aluminum production beyond the Deville process faced three key problems. The first challenge was to select an electrolyte containing an aluminum compound that could have electric current passed through it in an appropriate bath at a reasonable temperature. In general, two kinds of electrolytes are possible: halides (salts of fluorine, bromine, chlorine, or iodine) and oxides (salts of oxygen). Deville had selected a bichloride ($AlCl_3$, NaCl), whereas the process patented by the Swiss inventor Edward Kleiner-Fiertz was based on a bifluoride (derived from the rare mineral cryolite AlF_3, 3NaF). The second problem was to optimize the electrolytic bath by using a substance with a low melting point and a low electrical resistance, to minimize power consumption. By its composition and density, the bath should also ensure that the electrolyte could be easily dissolved and the metal produced easily isolated and extracted. The third issue was the cell technology, a particularly critical one when it came to industrial scale.[4]

Chemists in the United States and Europe were abuzz with ways to overcome these challenges and created a competitive atmosphere around this challenge not unlike today's X Prize or Breakthrough Prize. Professors shared this challenge with their students, and many young chemical entrepreneurs began to consider options. During the years 1874 and 1875, a young chemist named Frank Jewett went on a sabbatical to Europe and spent time at the University of Gottingen in Germany, where he met the pioneering isolator of aluminum metal Friedrich Wöhler. This visit was to be momentous in the history of industrial development, for it sparked the motivation in Jewett to focus his research and teaching around the chemistry of aluminum. After teaching for a year at Harvard as an assistant to Professor Oliver Wolcott Gibbs, one of the leading metals chemists of his time, Jewett was sent to Japan to teach chemistry at the Imperial University of Tokyo in 1876. Jewett gained further insights into metal-refining techniques while in the East. In 1880, he married a young American hygienist, Sarah Jewett, in Yokohama. Sarah shared his love of scientific learning while also considering the health implications of material choices. The couple returned to the United States later that year and made their home base at Oberlin College in Ohio, where Frank got a job as a professor of chemistry. Oberlin had distinguished itself as the first coeducational institution of higher learning in America, and this progressive streak attracted the Jewetts.

As he later recalled during a class reunion at Yale, Jewett had bought a small piece of aluminum during his travels and showed it to his students at Oberlin one day, with a bold challenge: "If anyone should invent a process by which aluminum could be made on a commercial scale, not only would he be a benefactor to the world, but would also be able to lay up for himself a great fortune!" Turning to a classmate, one of his students, a precocious young chemist names Charles Hall, said, "I'll be that man." Charles was the son of a local pastor and lived only a short walk away from the Jewett residence. Under the mentorship of Frank Jewett, he converted the woodshed behind his home into a makeshift laboratory. In February 1886, only a few months after graduating from Oberlin, Charles started his quest to find a nonaqueous solvent capable of dissolving aluminum oxide. Aqueous solvents were unsuitable because they would only yield aluminum hydroxide. Learning from the work of Edward Kleiner-Fiertz, Charles settled on cryolite as the key mineral for his experiments. The mineral was rare, but it had a relatively low melting point of "only" 1,000°C. In comparison, pure aluminum melts at 660°C and requires .3 kWh to reach its melting point, which is why the energy required to recycle aluminum is far less than the energy required for primary ore production.[5]

With a homemade coal-fired furnace and bellows in his woodshed, Charles also tried to find a catalyst that would allow him to reduce aluminum with carbon at high temperatures:

> I tried mixtures of alumina and carbon with barium salts, with cryolite, and with carbonate of sodium, hoping to get a double reaction by which the result would be aluminum. I remember buying some metallic sodium and trying to reduce cryolite but obtained very poor results. I made some aluminum sulfide but found it very unpromising as a source of aluminum then as it has been ever since.[6]

With the assistance of his devoted sister Julia Brainerd Hall, Charles was finally successful in passing an electric current through a bath of alumina dissolved in cryolite. On a brisk winter morning in February 1886, a small puddle of aluminum formed in the bottom of the retort that the Hall siblings had been toiling with for days. Charles filed for a patent in 1886, entitled "Process of reducing aluminum from its fluoride salts by electrolysis," and the patent, numbered 400,664, was granted two years later. Historians such as Martha Moore Trescott would later wonder what role his sister Julia had played in the discovery. Had she been neglected because of her gender but deserved further credit?[7] Given the history of Oberlin College and its championing of women in science and education, as well as Charles's own devotion to his sister, such revisionism might not be warranted. In 1999, the great-granddaughter of Emily Brooks Hall, the younger sister of both Charles and Julia, put such speculation to rest in a letter to the magazine *Invention and Technology* in 1999 after reviewing family archives, oral histories, and letters, noting that the family "would not have wanted his [Charles's] accomplishments, or the efforts of his sister Julia to provide encouragement and family support, to be misinterpreted. Furthermore, it is my belief that Julia herself would be extremely distressed by this attempt to revise history."[8]

Where Julia's efforts were most consequential and deserve unquestionable accolades pertain to what happened once Charles Hall's discovery was contested by another young inventor across the Atlantic. The son of a tanner, Paul Héroult's interest in science was whetted by a lawyer family friend who gave the young Héroult access to his vast personal library. Héroult soon became fascinated with science, engineering, and inventions—and the mystery of aluminum, through Sainte-Claire Deville's book *Aluminum, Its Properties, Its Production and Applications*. In 1883, Héroult spent a year at the École de Mines de Paris, where he was mentored by various chemistry professors into taking up the aluminum challenge. Without finishing his coursework, he left university to focus on his passion. Mercurial, hedonistic, and temperamentally the polar opposite of his quiet and reclusive American codiscoverer, Paul Heroult was still just as focused on the quest for industrial aluminum. After the death of his father in 1883, Héroult converted the tannery into a laboratory for his experiments, while his mother gave him her life savings to buy a Bréguet dynamo to power his experiments. Paul was born the same year as Charles, and through a similar sense of competitive industrial discovery he arrived at almost an identical process of aluminum refining at around the same time.

Given such a remarkable coincidence of discovery, the two inventors fought a lengthy court battle over whose patent came first. Technically, Héroult had filed his patent first, but thanks to the meticulous records kept by Julia Hall, Charles was able to prove that he had made his discovery back in February 1886, ahead of Héroult. Ultimately the two inventors were able to settle their differences, and the discovery became known as the Hall-Héroult process. The discoverers were labeled the "aluminum twins."[9] Later in life, they became good friends and corresponded frequently. In 1997, when the site of Charles Hall's discovery was consecrated as a National Historical Chemical Landmark, the grandson of Paul Héroult, Bernard Guest, flew in from Paris to attend the ceremony. The serendipitous independent discovery of an electrolysis process to extract aluminum metal at low cost by two researchers on either side of the Atlantic is a remarkable tale of the convergence of human ingenuity.

With his patent confirmed, Charles Hall sought financial backing and headed up the Ohio River and across the state border to the industrial hub of Pittsburgh, Pennsylvania, where he met with the metallurgist Alfred Ephraim Hunt. While other investors, inebriated with the dominance of iron, had dismissed aluminum's potential, Hunt saw great things in this new metal. Hunt, along with George Hubbard Clapp, his partner at the Pittsburgh Testing Laboratory; W. S. Sample, his chief chemist; the head of the Carbon Steel Company, Howard Lash; the sales manager for the Carbon Steel Company, Robert Scott; and a mill superintendent for the Carnegie Steel Company named Millard Hunsiker, raised $20,000 (around $600,000 in 2022 dollars) and founded the Pittsburgh Reduction Company on Thanksgiving Day, 1888. The company started out with a small smelting plant in Pittsburgh, then established another plant on the outskirts of the city, in New Kensington, Pennsylvania. Given the need for cheap and dependable power, a third site was opened at Niagara Falls in 1895, and the company began making lightweight automobiles in 1901. Seven years later, the Pittsburgh Reduction Company became the Aluminum Company of America (Alcoa). Two years later, the U.S. government started investigating monopoly power at Alcoa for its controlling interest in up to 90 percent of the country's economically viable bauxite mines. Never in the history of industry had one metal risen from obscurity to such dominance. The meteoric success of Alcoa was no doubt a testament to the dedication of its early executives, but it was also emblematic of how industrial entrepreneurship around key commodities can concentrate wealth. How such wealth is then translated into longer-term sustainability often depends on the idiosyncrasies and personalities of the tycoons in question.

No other metal has seen such a rapid rise and fall in production costs alongside a burgeoning of a brand-new market of industrial products. Despite its apparent ubiquity, to be extracted economically aluminum must occur in at least around 30 percent of its ore. Economic geologists compare the grade in the ore needed for viable extraction with the overall abundance of the element and come up with the term called "concentration factor." This is a measure of how much more concentrated an elemental ore needs to be compared to its abundance in the earth's crust to be viably extracted. Thanks to the Hall-Héroult process for extracting aluminum from bauxite, for aluminum this factor is among the lowest of any element, as shown in table 4.1.

TABLE 4.1 Economic geology of metals

Element	Crustal abundance (%)	Minimum exploitable grade (%)	Concentration factor
Aluminum	8	30	3.75
Iron	5	25	5
Copper	0.005	0.4	80
Nickel	0.007	0.5	71
Zinc	0.007	4	571
Manganese	0.09	35	389
Tin	0.0002	0.5	2,500
Chromium	0.01	30	3,000
Lead	0.001	4	4,000
Gold	0.0000004	0.0001 (1 parts per million)	250

A deposit of aluminum needs to only contain between three and four times the average crustal abundance, that is, between 24 and 32 percent aluminum, to be economical. In comparison, a chromium deposit needs to be *three thousand* times more concentrated to be economical. Thus even small levels of metallic concentration for aluminum are adequate to start a mine. Industrial ingenuity and the persistence of chemists over a century finally paid off with a tipping point that heralded the age of aluminum. The Hall-Héroult process was pivotal in reaching that moment whereby the natural capital of the most abundant metal in Earth's crust could finally be extracted economically.

FORTUNE OF FOUNDRIES

The discoverers of the Hall-Héroult process both died within seven months of each other in 1914, just as the First World War was beginning and the demand for aluminum for the military-industrial complex was gaining strength on both sides of the Atlantic. It was a strange irony of fate that both men lived for almost the exact same time span. Both had amassed considerable wealth from their discoveries and subsequent careers within the relatively short fifty-one years of their lives. However, the culture of philanthropy in the United States, Hall's successful patent decision, and the establishment of Alcoa gave him a more formidable fortune. At his death, Charles Hall owned Alcoa shares worth $30 million, which would be around $650 million in today's currency. He was immensely devoted to the small college town of Oberlin, where he had gained his knowledge of chemistry and made his landmark discovery. He donated one-third of his estate to the college and also gave funds for the maintenance of a green space in the heart of the campus called Tappan Square. One-sixth of his estate was donated to Berea College in Kentucky, which was the first coeducational and racially integrated college in the southern United States. Thanks to the philanthropy of Hall and others, it has a $1.1 billion endowment in 2022, and all its 1,600 students attend without paying any tuition. It is one of a handful of U.S. colleges with no tuition for all admitted students from anywhere in the world and that only admits students in the lower 40 percent of the U.S. income quartile. Another one-sixth of his estate was to be donated to the American Missionary Association, recognizing the work of his family, particularly in Asia. Hall was an internationalist and made a provision in his will to give the remaining one-third of the funds to colleges in countries or regions of the world that he considered most promising for higher learning: Japan, continental Asia, Turkey, and the Balkan States in Europe. The trustees were tasked with selecting the institutions, and some funds went to establishing bridging programs as well. The Harvard-Yenching Institute was established through these funds and continues to be one of the leading centers of interdisciplinary Asian studies research. The trustees also gave funds to twenty-one other institutions in Albania, Bulgaria, Greece, Macedonia, Turkey, Lebanon, India, China, Japan, and Korea.

As I wandered around the Oberlin campus to undertake archival research for this book, Hall's philanthropic imprint was palpable. Despite his love for industry, Hall was deeply connected to nature as well. He also gave funds for the acquisition of seventy-seven acres of forested land to expand the town's arboretum, which today remains an important recreational space for the students and the wider community. Oberlin has duly recognized Hall's contributions in various spots on campus as well, and there is a walking tour that commemorates various sites of significance in his life. The Oberlin Heritage Center is housed in the former residence of Hall's mentor Frank Jewett. It is now a National Historical Landmark for the subsequent inspiration Frank Jewett provided to a generation of students across the gender divide. Within the heritage center, there is a replica of the woodshed that Charles and Julia used in making their discoveries. Hall's own home has a plaque noting its historic significance but is now used for student residency and is not open to the public. The chemistry department at Oberlin, which benefited most from Hall's reputation, sports a display at its entrance dedicated to Hall and his discoveries. There is a replica of the aluminum cap of the Washington Monument and a plaque with a metallic imprint of Hall's historic 1889 patent. Most memorably, there is a life-size statue of Charles Hall cast in aluminum, which was donated to the college by the banking tycoon Richard B. Mellon almost a hundred years ago. Richard, along with his brother Andrew W. Mellon, had made an astonishing investment in Alcoa's early years. This statue has become a cultural icon for the Oberlin community and is often dressed in colorful attire and symbols appropriate for the holidays or used as a site to post political placards and statements.

The wealth Hall generated from his landmark discovery to upscale aluminum had a characteristic impact on fortunes for generations to come in many directions. Such is the power of industrial production that its growth can generate myriad opportunities. The energy and materials that charge our primary industries convert that natural capital into financial capital, which in turn can generate broader "multiplier effects" of livelihoods and employment. However, the concentration of wealth among the founders and their early investors has become one of the key points of contention in contemporary capitalism. Aluminum's rise is emblematic of what can happen with a high-risk–high-reward investment opportunity. While Hall had no children and donated much of his fortune to well-considered philanthropic endeavors, the Mellons, who made a major portion of fortune from their Alcoa investments, passed on much of their wealth to their progeny. Timothy Mellon, the grandson of Robert B. Mellon, has a net worth of over a billion dollars and has been one of the most significant donors to the U.S.-Mexican border wall project initiated by former U.S. president Donald Trump. The philanthropy that can be traced to industrial empires is often highly capricious.

Among the other major philanthropists to arise from the Alcoa fortune was the Napoleonic heir to Hall's management of the company—Arthur Vining Davis. After graduating with honors from Amherst College in 1888, Davis joined the Pittsburgh

Reduction Company as a shop helper and bookkeeper but through his dedication gained the trust of Charles Hall. He rose through the ranks of the company and eventually became its chairman and oversaw much of its expansion. Like Hall, he remained single throughout his life and amassed a fortune, which at his death in 1967 at the age of ninety-five was estimated at around half a billion dollars in (over $ 4 billion in 2022 dollars). Davis linked his philanthropy to a vision for sustainable development of key communities where he felt industry had a responsibility to act. Most notably, he left a trust for a small town on the Saguenay River in Quebec called Arvida, which had been named after his initials (ArViDa). It was known as "the City Built in 135 Days" by Alcoa to establish a smelter that could be powered by hydroelectricity from the nearby river. It was described by the *New York Times* as a "model town for working families" on "a North Canada steppe." The site was also chosen because the soil of this region contained anorthosite.

Davis was keen to show that an industrial boom town could be sustainable. He recruited a New York planner to develop a company working-class town and a model metropolis 240 kilometers north of Quebec City. Around this same period, the interwar years, Henry Ford set out to create the ideal factory town in the Brazilian Amazon, naming it after himself as Fordlandia, but unlike Arvida, the project failed and was abandoned. The vision of Davis, however, flourishes to this day. The town was nominated by the province of Quebec to be listed as a World Heritage Site by UNESCO in 2011.[10] The neat "workmen's" homes were built on spacious lots along curving streets rather than on a utilitarian grid. Each house had distinctive architecture and a garden, to defy the industrial impulse for uniformity. Schools, churches, and community centers with free Saturday night dances for the workers were part of the development plan put forward by Davis. In its heritage-listing submission, the Quebec town of Saguenay, which now incorporates Arvida, described the sustainable characteristics of its urbanism as follows:

> A masterpiece of urban planning showcasing human creative genius, Arvida deliberately combines Western ideals and American legacy in a unique materialization of secular utopias. Remarkable for the holistic nature and symbolic dimensions of its plan, it is exceptional in its intention and scope, egalitarian housing, vernacular aesthetics, realization, and conservation.[11]

As the largest aluminum production center in the Western world, Arvida was so critical to the Allied effort during the Second World War that the wartime city was guarded by antiaircraft batteries.[12] In its heyday, Arvida had about twenty thousand residents; it still boasts around 12,000 residents and a stable real estate market. The smelter and other related plants, now part of Rio Tinto Alcan, are still in operation. Although the Canadian government decided not to submit Arvida's candidacy to UNESCO for a world heritage listing, Quebec declared it a provincial heritage site. Despite the politics of a largely Anglophone enclave in a fiercely Francophone region, the socioindustrial utopian vision of Arvida has transcended such parochialism. The town's fortune did contribute to what is often called the "Quiet Revolution of Quebec," wherein the province accumulated wealth and concomitant welfare programs that fueled some resource nationalism (or separatism), but Arvida and its surrounding region remained relatively calm during much of the tumult around separatist referenda. The economic historian John Hartwick, who grew up in Arvida, wrote a book on the town's unique features. One resident of Arvida, Cheryl E. Reid, wrote of her childhood years in the town in a letter to the *Globe and Mail* newspaper in Toronto:

> As a teen growing up in Arvida, the town provided a great sense of belonging. The community was vibrant, tight-knit, well-educated and adaptable to living in this remote region of Quebec. The English-speaking children were fortunate to acquire French while very young. English TV didn't make its way to the region until 1976. The streets of Arvida were appropriately named Joule, Watt, Ampere, Lavoisier, Gay-Lussac and Coulomb after units of energy or world-renowned scientists. The first aluminum bridge in the world was built there in 1950, crossing the Saguenay gorge near the Shipshaw powerhouse.[13]

Arvida symbolizes what a visionary industrialist could achieve with careful planning and long-term financial investments in a sustainable future for a community. The proverbial tale of industrial "ghost towns" animates many dystopian narratives about factory-fueled human settlements. Yet industry can also be just as much a spur for sustainable community planning. The conversion of natural capital to financial capital that happened in Arvida could have backfired into a downward "bust." There are myriad examples of communities, provinces, and nations afflicted with what is often called the "resource curse" by economists. The endurance of aluminum's uses and the continuation of the smelter in the town has certainly helped in maintaining the town's economy. However, it was careful planning that ultimately succeeded in making sure that the town diversified its economy. The city of Saguenay, of which Arvida is now part, has over 145,000 people. All aluminum-related industries now comprise around thirty thousand jobs, and the goal is to leverage the town's knowledge base in metallurgy and hydropower for a broader long-term viability of employment. There are limits, though, to what good planning and philanthropy can achieve. Further north of Saguenay is another town that also owed its origins to the Hall-Héroult process but that had a very different fate.

THE ARCTIC MINE AND ITS SYNTHETIC SUNSET

The "secret sauce" of the Hall-Héroult process for upscaling aluminum production was the rare mineral cryolite. The more chemically accurate name, one that denotes its constituent elements, was "sodium hexafluoroaluminate." In its molecular composition, a central aluminum atom is surrounded by six fluorine atoms, giving three units of negative charge to the anion. This assemblage is electrochemically embraced by three sodium cations with a single positive charge each. It is that crystalline arrangement of atoms that gives cryolite its relatively low melting point, offering the possibility of coaxing aluminum metal out from alumina in an energy-efficient manner. Cryolite consists of 12.85 percent aluminum, 54.30 percent fluorine, and 32.85 percent sodium. Since its refractive index is approximately that of water, transparent, colorless cryolite becomes almost invisible when placed in water. The ice-like appearance of the mineral is the reason why its name was derived from the Greek words *kryos*, or "ice," and *lithos*, or "stone." Even though aluminum minerals are ubiquitous on Earth's crust, cryolite was particularly rare because its formation required specific geological conditions that had only occurred in a few spots. Some geologists thought that cryolite formed via the action of fluoriferous gases upon the original granite magma where the mineral formed; others proposed that it formed in quartz-deficient regions with a very slowly melting magma flow. Such conditions have led to cryolite formation in Siberia, West Africa, and the Americas, but in very small quantities. The only known location where these geologic conditions occurred as a persistent feature for long enough in geological time to produce an economically viable deposit for extraction is the southwestern coast of Greenland.

It was in Greenland that cryolite was first described and named in 1798 by the Danish veterinarian and physician Peder Christian Abildgaard (1740–1801), who had a passion for natural history and made expeditions to the territory. He had been alerted to the discovery of an unusual stone in the shipments of mineral specimens that arrived regularly with seal fur and other local merchandise in ships of the Royal Greenland Trading Company. Greenland's indigenous Inuit people called this hitherto unclassified stone "the ice that never melts" in their language and recognized its anomalous properties compared with the parent granite in which it was embedded. Another local name for the mineral was *Orsuksiksa~t*, meaning "the stone looking like the seal's blubber." The Greenlanders used it for ornaments and even crushed it and used the powder with tobacco for snuff. For the first fifty years of its discovery, cryolite was merely a curiosity for museum collectors and some experimental chemists who tinkered with its specimens once its elemental composition was confirmed by the great chemist Berzelius around 1823. There were attempts made to use it as a glaze for porcelain, which failed. Since it was a sodium mineral, the most promising prize for its chemistry in those times would be for the production of sodium carbonate, or washing soda. As its common name suggests, this chemical was the key ingredient in household soap; it was also used in glass manufacturing.

Sodium carbonate had previously been extracted from processes that had serious drawbacks. During the eighteenth century, the earliest industrial manufacture of the chemical came from a surprising botanical source—coastal seaweed and kelp forests in Scotland were burned to produce soda "ash" (hence the origin of the term). However, around twenty-three tons of seaweed was required to produce one ton of "kelp ash." The kelp ash would consist of around 5 percent sodium carbonate. In 1775, the French Academy of Sciences held a public competition to find a way to manufacture soda ash. This competition went on for years until in 1791 the French surgeon Nicolas Leblanc devised a process involving simple salt and sulfuric acid and subsequent reactions with limestone-derived calcium carbonate. Before he could claim this prize money, the French Revolutionaries seized his operation and leaked his trade secrets.

A decade later, after Napoleon's ascent to the throne, the plant was returned to Leblanc, but without his prize money; the inventor sank into despair and committed suicide in 1806. His process took hold in Europe and the United Kingdom but had a serious drawback—the production of calcium sulfide as a byproduct, which had no commercial use. For every eight tons of soda ash, the process produced 5.5 tons of hydrogen chloride and seven tons of a waste material, which acquired the onomatopoeic name *galligu*, for its putrid, viscous composition. Galligu largely comprised calcium sulfide but also contained a range of heavy metals and arsenic. Since it had no economic value, it was discharged onto fields near the soda works, where it would undergo further reactions in the wet organic soils and release hydrogen sulfide—the toxic gas responsible for the odor in rotten eggs.[14]

The public nuisance caused by the spread of galligu was so acute that it galvanized the first major environmental litigation actions in the United Kingdom. One 1839 lawsuit against soda works alleged:

The gas from these manufactories is of such a deleterious nature as to blight everything within its influence and is alike baneful to health and property. The herbage of the fields in their vicinity is scorched, the gardens neither yield fruit nor vegetables; many flourishing trees have lately become rotten naked sticks. Cattle and poultry droop and pine away. It tarnishes the furniture in our houses, and when we are exposed to it, which is of frequent occurrence, we are afflicted with coughs and pains in the head . . . all of which we attribute to the Alkali works.

Among the earliest environmental laws ever passed in the United Kingdom were linked to such pollution. This 1863 Alkali Act and several subsequent sets of "Alkali Acts"[15] regulated both the production of hydrochloric acid and galligu. To comply with the legislation, among the earliest pollution-control technologies were devised. Soda works passed the hydrogen chloride gas emissions through a charcoal-packed tower, where flowing water helped absorb the acidic fumes and prevent them from escaping. Pollution and waste control thus motivated the innovations that brought cryolite to industrial attention.

After years of efforts, the chemist Julius Thomsen patented a method to separate cryolite into soda in 1853. The process did not generate any acidic pollution and was economically quite efficient. His reaction essentially involved reacting quicklime with cryolite under specific conditions. There was no pollution concern from the reaction, and it was economically efficient. Given the scale of the soda market, this patent made it economically viable for a mining site to be set up in Greenland. The entire amount of available cryolite at the time in Denmark was a few hundred kilograms of specimens from the Royal Danish Trading Company. The geologist and explorer Hinrich Rink was dispatched to Greenland by the Danish government to scope out the possibility of a mine being set up near the Inuit settlement of Ivigtut. After his positive response, Thomsen and his associates were able to find a few investors with interests in commerce in the northern seas, prompted by the subsea telegraph cable lines being set up, which heralded a new era of transatlantic communication and business ventures. A small mining town was set up within a couple of years, and the first major shipment of cryolite from the only mine of its kind arrived in Copenhagen in 1856.

The soda manufacturing use of cryolite was relatively short-lived: the Belgian chemist Ernest Solvay discovered yet another process to make sodium carbonate using ammonia, for which no exotic mineral was required. However, as the use of cryolite for soda manufacturing began to lose market share to the Solvay process (which is still in use today), the Hall-Héroult process picked up momentum toward the end of the nineteenth century. By 1900, the mine comprised an expanding open pit, dormitory-style accommodation, machine shops, loading docks, a medical clinic, and a 175-man workforce. Even though the Ivigtut region still holds the record for the highest recorded temperature in Greenland (30.1 degrees Celsius), it was still a physically demanding location for a mine. Given the harshness of Greenland's climate, mining was only possible in the summer, and miners flooded the pit with seawater in the autumn to prevent it from filling with snow. In spring, they would break the ice and pump the water out to restart quarrying. It was only between the months of April and October, when the sea was clear of ice, that ships could navigate the waters to the port and load the cryolite for delivery to customers in Canada, Europe, and, most significantly, the United States. The Pennsylvania Salt Manufacturing Company, or Pennsalt, was the largest customer for the cryolite mined during the 1930s and 1940s; it in turn sold the mineral to Alcoa for use in aluminum manufacturing.

FIGURE 4.2 The Greenland cryolite mining town of Ivigtut
(now known as Ivittiit) in 1940.

Source: Wikimedia Commons.

As the demand for aluminum burgeoned and the fortunes of
Alcoa grew, there was increasing anxiety about the dependence
of the production process on this one single mineral source in a
remote, isolated location. Researchers at the company were tire-
lessly working to find a way to synthesize cryolite from more
abundant primary materials. Finally in 1937, three researchers at
Alcoa led by John Morrow filed a successful patent for the synthesis of cryolite from fluorospar (fluorite, or calcium fluoride)
and sodium aluminate. Although fluorspar still needed to be
mined, it was far more abundant and accessible, as was sodium
aluminate. The availability of synthetic cryolite (sodium aluminum fluoride) began to reduce the demand for natural cryolite
toward the late 1930s. However, during World War II, the critical need for aluminum demanded huge quantities of both natural and synthetic cryolite. The U.S. government thus built a navy
base near the Greenland cryolite mine called Green Valley. During the initial period of U.S. humanitarian aid to Greenland
(1940), the cryolite mine was identified as the only sensitive military target in need of protection. In addition to fears of German

attacks, there was also concern about labor unrest at the mine, and to allay these concerns a force of 15 U.S. servicemen were given discharges and then hired by the Cryolite Company. This wartime town, Kangilinnguit, is now a Danish military base. It is connected by a five-kilometer road to Ivigtut, which was subsequently renamed "Ivittuut." The production of the mine peaked in 1942, when it mined and shipped 85,000 tons of cryolite to North American aluminum smelters.

In December 2019, the *Smithsonian* journalist Katie Lockhart published an article about her visit to Ivittuut, which has now become a regular stop-off for Greenland tourist cruises. Her interview with Rie Oldenburg, a historian and head of education at Campus Kujalleq, a school in nearby Qaqortoq, Greenland, revealed that given security concerns, no photographs of Ivittuut were allowed to be taken during the war, and no one was allowed to write letters to family or friends for fear that the Germans would intercept them. Her visit took place during the presidency of Donald Trump, and when she asked local Inuit about what it was like during the war and their relationship with Americans, they declined to answer, fearful that their responses would reflect poorly on the United States at a time when the president was openly mulling purchasing the island. The locals did indicate that American soldiers had left a remarkable testament to materialism behind that would shape modern Greenland consumer culture—Sears, Roebuck and Company catalogs! Long before the ubiquity of Amazon Prime, these catalogs allowed Inuits and Danish Greenlanders alike "to order name brand appliances, like General Electric stoves and refrigerators and boats that modernized the way of life in Greenland."[16]

The mine continued to operate with a 1,500-foot-long inclined tunnel and workings as deep as two hundred feet below sea level. The world's only economically viable cryolite mine had produced

3.7 million tons of ore grading 58 percent cryolite by 1962, when it was officially declared "depleted." Mining operations ceased, and only small crews remained to clean up the old dumps. By 1987, the town of Ivittuut was abandoned, and its infrastructure and quarrying cavity became heritage sites.

But the confluence of geology and technology should never be underestimated—even in Cape Desolation, Greenland. Fast forward to the year 2021: mineral entrepreneurs from a distant land—Australia—acquired the mineral rights to the ore deposits around the Ivittuut mine. A Per-based company called Eclipse Metals recognized that in addition to cryolite, the mine workings contained associated minerals including fluorite, siderite, quartz (high-purity silica), rare-earth elements, and base metals. With a focus on minerals required for the new technologies of the green-energy transition, the site could yield a profitable resource for the company. Carlos (Carl) Poppal, the company's CEO, has noted that the large tenement area covers Greenland's only known carbonatite deposit, which can provide an ideal product for neutralizing acidic mine and process water produced by other miners in the region: "This fits well with the company's mission to excel in the commercialization of metals and minerals demanded in the production of green energy and required by the industry in the reduction of pollutants."[17] The Greenland government has also encouraged the reopening of the mine, and the exploration license has been extended to at least 2024.

The deep, complex, and dramatic stories of mineral availability and supply are inextricably linked to the history of human livelihoods and security. Cryolite became essential for aluminum production merely by acting as a solvent for aluminum's primary ore and making it cost-effective to extract the metal by lowering the temperature, and hence the energy demand, for the reaction. It was that specific innovation of the Hall-Héroult process that led to the development of a small Inuit village in the Arctic and even an American military base. Minerals as primary inputs to industry can muster such marvelous levels of influence that they can drive and transform the lives of individuals, communities, and countries.

III

FLIGHT AND FOIL

This valuable metal possesses the whiteness of silver, the indestructibility of gold, the tenacity of iron, the fusibility of copper, the lightness of glass. It is easily wrought, is very widely distributed, forming the base of most of the rocks, is three times lighter than iron, and seems to have been created for the express purpose of furnishing us with the material for our projectile.

—Jules Verne, *From the Earth to the Moon*

Relentlessly, we must guard these vital lifelines of supply
Without bauxite we have no aluminum
Without aluminum we have no wings
Without wings we have no defense

—Narration in U.S. government video on protecting
aluminum supply chains, 1940s

5

MOBILE METAL

How Aluminum Facilitated War and Peace

Our contemporary culture of innovation and entrepreneurship remains deeply entwined with the military–industrial complex.
—Mimi Sheller, *Aluminum Dreams*

I n a celebratory issue of *Science* published in 1936 to commemorate the golden jubilee of the Hall-Héroult process, Arthur Vining Davis was asked about Alcoa's industrial strategy. At that time, Davis, who had served the company his entire career, was lauded as one of the most astute business executives in America. He presided over a vast material empire with unprecedented vertical integration, from mines to consumer markets. Alcoa thus symbolized the success of American chemical innovation and business enterprise. In this spirit, Vining described the four "epochs" of the company to the editors of *Science* as follows:

(1) Can we make aluminum? And this we were able to answer in the affirmative as soon as our production reached 30 pounds per day. (2) What can we do with what is made? Which became an early problem as our output, small as it was, piled up on our hands

and was answered by making novelties of it. (3) Can we make any
money on it? This was finally answered by our going into the business of doing our own fabricating. (4) How can we make the
business grow? This keeps us searching for new materials through
research, despite the fact that our present production is in the
neighborhood of 300,000,000 pounds per year (1936).[1]

The commercial production of aluminum happened at a time
of great global unrest, and as with many other scientific endeavors, the material was co-opted by the military-industrial complex. The metal's light weight, coupled with the abundance of
its ore, made it an ideal component of transport infrastructure.
This chapter considers the role of aluminum companies in working with aircraft manufacturers, particularly during World War II,
and the securitization of metal production. The ecological toll of
such securitization of materials provides broader insights about
resource efficiency when defense imperatives trump environmental concerns. Metals have always been synonymous with security, from the armor of knights to the bullets and bomb casings
of drone missiles today.

While the label of "metal" is not inherently definitive at the
atomic level, humanity has generally tagged elements with certain key physical properties of strength with this nomenclature.
Strength and malleability are linked to the chemical bonds that
exist between individual metallic atoms, and in general the substances we call metals have very strong bonds and thus high
melting and boiling points. However, there are anomalies in this
regard as well—particularly zinc, cadmium, and mercury.
Because the electronic configuration of these three metals is
more stable, the formation of adjoining bonds between metallic
atoms is somewhat impaired. They can thus volatilize more
easily—mercury being the most extreme example of this, whereby
it can turn to vapor even with gentle heating. Human ingenuity

has made use of almost every distinctive physical property of an element. Mercury's peculiar volatility and its ability to form temporary bonding lattices with other metals (called amalgams) are used to extract gold from diffuse sediments.

The other metal that can be a liquid at room temperature—gallium—presents a nuanced variation on bond strength. While a wad of gallium can famously melt in the palm of your hand, the actual bond strength within the metal is itself quite strong, and its boiling point is comparable to a metal like copper. Why, then, is gallium easily liquefied but not vaporized? The answer is because bond strength does not always require a rigid lattice structure. Bonds can be strong but pliable. As humanity wrestled with ways of manufacturing optimal materials for myriad new industrial purposes, such anomalous properties became ever more valuable. Slight variations in electron orbitals can produce elemental properties that defy the periodic table's attempts at imposing a fixed nomenclature. Gallium and aluminum are neighbors on the periodic table and, in fact, gallium's heavier weight would have us think it would have more "metallic" properties. Yet its conductivity, metrics of physical strength, and chemical behavior make it far more ambiguous in its properties.

The price of metallic strength in the material realm has often been physical weight. Lead is strong and dense enough to shield us from highly penetrative gamma rays, but it is proverbially heavy. Steel is synonymous with strength, which makes it so favored in construction. Some of the categories in which a metal's strength can be demarcated have been defined by metallurgists as follows:

- *Tensile strength* connotes the maximum amount of pulling and stretching a metal can endure.
- *Yield strength* is the stress point at which metal begins to deform plastically.

- *Ultimate strength* describes the maximum amount of stress a metal can endure.
- *Breakable strength* is the stress coordinate on the stress-strain curve at the point of failure.
- *Impact strength* is a measure of how much impact or rapidly applied force a metal can take before it fails.
- *Compressive strength* is the maximum amount of pressure or compression a metal can withstand.

The impact load and the limit that a metal can withstand are expressed in terms of the amount of energy that a metal can absorb before it fractures. This is typically measured with a universal testing machine that applies an increased load to the material.

In addition to these measures of strength, metals are also defined in terms of *malleability*, which connotes the potential to deform under pressure (compressive stress). Malleable materials may be flattened into thin sheets by hammering or rolling. A similar property, that of the metal being able to be drawn into a wire, is referred to as *ductility*. Such a property is only of use, however, if the metal does not lose too much strength while being stretched. Ductility and malleability are not always correlated; as an example, gold is ductile and malleable; lead is merely malleable. Such a variety of physical properties of metals are caused by the variances in their crystalline structures. Metals tend to fracture at "grain boundaries," where their atoms are not as closely bound by electrochemical forces. Most metals become more malleable with increased temperature, as temperature affects the crystal grains. No metal is completely malleable, and all will break at some level of stress. Ductility is in many ways a form of tensile plasticity. Molecular-level forces and crystal-grain gaps are collectively involved in allowing for deformation while maintaining the integrity of a metal. Functionally, ductile properties are important for preventing the catastrophic failure of structural components during service, whereas plasticity is critical for shaping and forming metals into the desired shape and geometry to make structural components.

The underappreciated polymath Robert Hooke—termed by some commentators as the "English Leonardo"—is credited with first presenting a practically useful synthesis of such key functional properties of metals. Following the Great Fire of London in 1666, Hooke was asked by the great architect Sir Christopher Wren to help with rebuilding thousands of buildings. As he was studying the contortions of metals during the fire, he discovered the physical "law" that still bears his name. Hooke's Law states that the force required to stretch material is proportionate to the extension of the material. When he discovered this law of elasticity in 1676, he didn't publish it in the ordinary way. Instead, he published it as an anagram: "ceiiinossssttuv." Such cryptic publication of discoveries was common in those days to maintain intellectual property. This ensured that if someone else made the same discovery, Hooke could reveal the anagram and claim priority, thus buying time in which he alone could build upon the discovery. He revealed the solution to his anagram two years later as the Latin phrase *ut tensio, sic vis*, meaning "as the extension, so the force."

Figure 5.1 shows the relationship between two key properties linked to Hooke's Law—stress and strain. Although these terms are often used interchangeably in common English, there is a nuanced difference between the two in physics. Stress is defined as the deformation force per unit area of the body or material. Strain is the ratio of the amount of deformation experienced by the body in the direction of force applied to the initial sizes of the body.

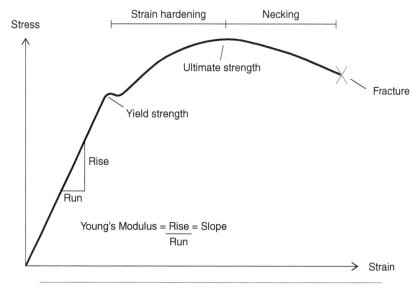

FIGURE 5.1 Metal strength metrics.

Source: Adapted from Autodesk.com.

The shape of this graph is crucial in determining why certain metals are well suited for construction and other high-intensity uses where they come under stress and experience strain. The linear left portion of the graph is where the material is going through "elastic" deformation and will return to its original position when the stress is reduced—this is the Hooke's Law portion of the relationship. The bonding forces between atoms of a material here will behave like Hooke's experimental springs. The slope of the linear relationship in this graph signifies the rigidity of the material; it was formally codified by the English engineer Thomas Young in the nineteenth century and bears his name as Young's modulus. The steeper the curve, the more rigid the material. Ceramics have the highest Young's modulus, followed by metals and then a range of other more elastic materials. Beyond the "yield strength" point, the material goes through a "plastic deformation": it will not return to the original state but can still bear the stress without fracturing. Malleability and ductility fall in this extended range, and this is where the crystal-grains aspect of the material deformation dominates over any bonding forces between the atoms of the material.

However, it is not merely the strength of the metal that matters for weapons and the broader military-industrial complex—it is also the tradeoff with weight or its concomitant density. Figure 5.2 presents the comparative strength of a range of materials in relationship with their density. Metals generally are found

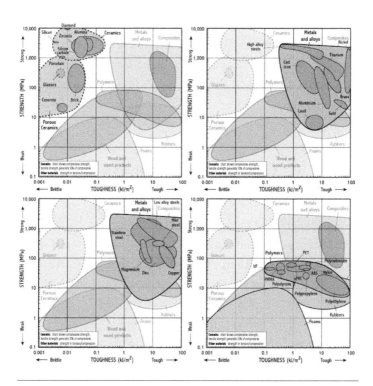

FIGURE 5.2 Comparative strength of a range of materials, with a focus on metals.

Source: Figures from University of Cambridge, Department of Material Science and Metallurgy, used with permission.

in the northeastern quadrant, with high strength and density. Aluminum is particularly versatile, with a relatively low weight but a range of strength prospects based on its alloying materials.

Oscar and Wilbur Wright recognized this sweet spot of aluminum between its light weight and its strength. Even though in 1903 the metal was expensive, they chose it for the central mechanical part of their plane—the engine. This pioneering machine had four horizontal inline cylinders with a four-inch bore and four-inch stroke. While these were made from cast iron, they were placed inside a cast aluminum crankcase that created a water jacket around the cylinder barrels. By using aluminum, they ensured their engine would be light enough to get the plane off the ground. As Jules Verne had predicted, aluminum would be the metal to help humanity take flight. Reaching for the sky with our metallic machines spurred human ingenuity—for better and for worse.

VICE AND VIRTUE OF
METALLIC VESSELS

Rummaging through the archives of the Allegheny Valley Heritage Museum in Tarentum, Pennsylvania, I came across a small book published by Alcoa in 1943 titled *Our Metal in War and Peace*. The book was written at the height of World War II and declared that "aluminum is a vital war material . . . and when Peace arrives, our metal will be released from military duty, free to resume its civilian affairs, along with the thousands of Alcoa Family men and women in uniform." There was also a small article from the journal *Nature* dating back to February 11, 1915, titled "Metals and War," which enumerated the various metallic needs of Germany. The article concluded that even a supply curtailment of one minor metal like manganese—which is essential for steel—could affect its war effort. The criticality of metals from the age of knights to the twentieth century's mechanized battles was salient as ever.

Napoleon III had patronized aluminum for defense purposes, and France made the first recorded military use of aluminum as far back as 1892, in torpedo boats. During the Spanish-American War, just a few years later, the United States used aluminum in tent poles and canteens; future president Teddy Roosevelt carried an aluminum canteen during that war. A history of this period notes that "United States cavalrymen fighting in the Spanish-American War tethered their horses to aluminum picket pins, and infantry troops slept in tents pegged to aluminum stakes."[2] The first military aviation use of aluminum was likely in the early aircraft of the world wars, especially World War II. Aluminum fuselages and other critical parts, like landing gear and propellers, were lightweight while retaining their strength. This was a time of remarkably accelerated innovation in aircraft engineering, and aluminum can be credited for spurring optimism in our technical battle to vanquish gravity.

In World War II, air supremacy was a key strategic advantage, and metal supply chains for aircraft thus were a crucial security feature. The Alcoa plant at Arvida became the powerhouse for the Allied war effort by supplying most of the aluminum for the planes that would ultimately be consequential in winning the war. Arvida became a "secret city" protected by an extensive contingent of military special forces. The Commonwealth of Canada declared aluminum a "wartime industry," and the city was protected by the highest level of military defense in the country. The Bagotville military base was constructed nearby, and the U.S.-Canadian military alliance around aluminum became pivotally consequential. An American metals company was producing aluminum on Canadian soil for a war effort that spanned the globe. During both world wars, about 90 percent of aluminum production in North America likely went into military uses; the metal remains essential to many components of modern warfare. As the historian George David has argued in his meticulous history of Alcoa, such an immediate need for aluminum likely also led to the company's dominance and monopoly power. There was a need for efficient mobilization of metal, and working with one supplier was easier for governments, even though such singular reliance was also problematic.[3]

The sociologist Mimi Sheller has documented the ways that aluminum, the military, and the political economy of innovation became so sacrosanct at the expense of many wider communities from where the metal was sourced. She states that "our contemporary culture of innovation and entrepreneurship remains deeply entwined with the military-industrial complex, with serious implications for our ability to address ethical issues concerning global pollution, environmental destruction, and the huge effects of aluminum production on marginalized people."[4] She goes on to document how aluminum made possible a range of malevolent innovations, including the infamous "daisy-cutter bomb" BLU-82, which was used in Vietnam and more recently in the wars in Iraq and Afghanistan. The bomb, which is often called "the world's largest non-nuclear weapon," contains a slurry of ammonium nitrate, aluminum powder, and a polystyrene-based thickener, which when it explodes generates a massive pressure wave estimated at 1,000 pounds per square inch. During World War II, Britain's Royal Air Force used bundles of thousands of strips of coarse black paper with aluminum foil stuck on one side as a radar-jamming mechanism. The Royal Air Force is believed to have won the aerial battle for Hamburg largely thanks to over seven thousand of these bundles dropped on the city.

Usage of aluminum for military purposes continued to grow substantially in the middle of the twentieth century, and there were even points in time when restrictions were placed on the use of aluminum for nonmilitary purposes in the United States. A stockpile of the metal began to be kept by the government during the Korean War and continued all the way up to the Vietnam War period. However, the rise in production of the metal also increased substantially during this period, and eventually the stockpile became unnecessary. In 1965, the Council of Economic Advisors was concerned about the stockpile causing a wartime inflationary problem in heavy industry across the country and suggested some release of the metal into the market as well as buyback programs by industry. This posed an intriguing conundrum for the industry: reduce growth potential through the buyback program but establish a more favorable relationship with government. Although environmental considerations were hardly on the horizon at the time, the buyback program, by relaxing production, would also relieve energy usage, mining impacts, and air pollution from smelters. However, potential job losses might follow from reduced production. After tough negotiations, a settlement was reached for an orderly release of the stockpile over seven to fourteen years (depending on natural fluctuations in demand). The role of aluminum in winning the world wars was not lost on the marketing department at Alcoa. Figure 5.3 shows an advertisement widely distributed across the nation at that time, one of the early examples of an "infomercial." The director of the U.S. Public Affairs Institute authored a report in 1951 titled "Aluminum for Defense and Prosperity" and noted that the metal had "become the most important single bulk material of modern warfare. No fighting is possible, and no war can be carried to a successful conclusion today, without using and destroying vast quantities of aluminum."[5]

THE TIMETABLE OF ALUMINUM FOR VICTORY

1938 Sept. *Munich.*
Oct. *Czechoslovakia invaded.*
Nov. Alcoa inaugurated an expansion program which by 1943 cost $250,000,000.
Dec. Alcoa 1938 production, 287 million pounds; started 1939 with more than a year's supply on hand.

1939 Jan. Alcoa begins operation of new extrusion and tube mill.
Feb. Alcoa starts building excess stock pile of airplane sheet.
Apr. *Congress authorizes Army to acquire 6,000 planes in 2 years; and Navy 3,000 in 5 years: Approximately one month's 1943 goals.*
Sept. *Poland invaded.*
Alcoa authorizes new metal-producing capacity.
Nov. *Finland invaded; Cash-and-Carry Act signed.*
Dec. Alcoa authorizes another huge metal-producing plant, although it begins new year with 215 million pounds on hand.

1940 Jan. *First Request for defense appropriation in Budget Message.*
Mar. Alcoa reduces price of ingot from 20c to 19c.
Apr. *Denmark, Norway invaded.*
May *Low countries invaded.*
First of new Alcoa metal-producing plants starts production.
June *Dunkerque; France capitulates.*
Alcoa authorizes still another metal-producing unit.
July *Congress gives first go-ahead on faster plane production.*
Aug. Alcoa reduces ingot price from 19c to 18c; adds large alumina capacity.
Sept. *Egypt invaded; Selective Service Bill passed.*
Another new plant starts operating; still more units authorized.
Oct. *Rumania invaded.*
Nov. Alcoa reduces ingot from 18c to 17c; authorizes still more metal-producing capacity.
Dec. Alcoa faces new year with 154 million pounds on hand.

1941 Jan. *NDAC says aluminum supply adequate to meet October, 1940, estimates of requirements.*
Feb. *Aluminum put on priorities to give all capacity to defense.*
Mar. *Lend-lease.*
Apr. *Yugoslavia invaded; U. S. occupies Greenland.*
May Very large new Alcoa metal-producing units start operation.
June *Crete lost; Russia invaded.*
Alcoa authorizes further expansion at own expense.
July *Government authorizes first of its own plants to supplement the enormous expansion of Alcoa.*
Aug. *Government announces Alcoa will build and operate 3 of these plants. (All Alcoa designing and building of Government plants done without profit.)*
Sept. *Government decides on more plants; instructs Alcoa to build them.*
Oct. *Government decides to build more plants; instructs Alcoa to build them.*
Alcoa reduces ingot from 17c to 15c.
Nov. *Government reviews detailed plans for own large sheet mills.*
Dec. *Pearl Harbor; Churchill-Roosevelt strategy conference.*
First metal rolled on Alcoa's own 50-times-faster sheet mill, largest in the world. Alcoa receives further new plant instructions from Government.

1942 Jan. *Pan-American Conference.*
Another Alcoa metal-producing plant in operation; additional instructions from Government to build new plants.
Feb. *Government authorizes large alumina plant, several aluminum plants and large sheet mills. Alcoa to design and build.*
Mar. *Government authorizes Alcoa to build more casting capacity, extrusion capacity, and forging capacity.*
Apr. *Government authorizes blooming mill and enlarged tubing capacity.*
May *Government-owned plants, built and operated by Alcoa, start operation.*
June *Special authorization for airplane cylinder-head capacity.*
July to Dec. *{As instructed by Government, Alcoa starts additional plants for various types of fabrication, as special needs of war production are made apparent by changing emphasis on war equipment.*

1943 The aluminum industry will have a metal capacity of over two billion pounds, seven times prewar. Alcoa has more than doubled its metal producing and fabricating capacity through a self-financed expansion program. The expansion by private industry has been augmented by a vast Government program where the kind, amount and time of expansion has been at the direction of the Government. In addition to operating its own twenty plants, Alcoa has been honored with the responsibility for constructing and operating 40 Government projects in 25 different locations.

ALCOA ALUMINUM

FIGURE 5.3 Alcoa's nexus with the war effort was proudly advertised in this widely published 1943 timeline.

Source: Alcoa Archives.

The Axis Powers in World War II relied on much of their bauxite from France, Croatia, and Hungary. Natural gas supplies from Transylvania for smelting operations were also critical. Indeed, a key motivation for the Nazi regime's interest in the conquest of France was for access to its large bauxite deposits. Germany appointed an "aluminum trustee," the businessman Heinrich Koppenberg, who was able to secure supplies for the regime from a variety of sources and was particularly active in France for ore and Norway for processing capacity. Between 1940 and 1943, France exported 68 percent of its alumina and aluminum production to Germany. The occupation of Greece by the Axis Powers also gave them access to massive bauxite deposits. Before the war, American banks had funded the Hellenic Hydro-Electric and Metallurgical Corporation. When Greece was occupied, the U.S. government prepared a report in 1943 under the Metals and Minerals Branch of the Office of Economic Warfare detailing the locations of bauxite mines of use to the Allies after liberation of the country.[6] Both sides realized that winning any world war was predicated on the ability to transport personnel and supplies over long distances. Winston Churchill had presciently written in his 1899 book *The River of War*: "Victory is the beautiful bright-colored flower. Transport is the stem without which it would never have blossomed."

The industry was also investigating aluminum's peacetime uses. The Cold War initiated an episode of competitive science that also required aluminum. The Soviet Union and the United States were locked in a space race that required aluminum not just for the structure of spacecraft but for the fuel that would propel them into orbit. Solid rocket fuel can be traced back to the early fireworks developed by the Chinese centuries ago. Refined aluminum powder and the oxidizing mineral salt ammonium perchlorate would be instrumental in rocket fuel. This mixture, when packed together with an organic binder, polybutadiene acrylonitrile, or PBAN, has the consistency of a rubber eraser and can be packed into a steel case. The burning of this fuel produces aluminum oxide, aluminum chloride, water vapor, nitrogen gas, and a vast amount of energy, which can heat rocket boosters to more than 3,000°C, which leads to the rapid expansion of the water vapor and nitrogen. This expansion creates the massive thrust factor that propels a rocket forward.[7] There has been some concern about the residue from this fuel on stratospheric ozone depletion, detailed in research conducted by the chemist Darrel Day Spencer, a student of the Nobel laureate Mario Molina (who codiscovered the impact of chlorofluorocarbons on ozone).[8] However, the volume of space flights is still too small to have a large-scale impact. As space tourism and more satellite launches occur, however, such impacts will need to be monitored.

Once in space, materials need to withstand immense stresses, and aluminum alloys are immensely useful in this context as well. Any orbiting satellite or spacecraft that is constantly moving in and out of the sun's direct heat endures huge temperature fluxes, which can cause its components to expand and contract. A material's ability to maintain its size and shape connotes what is called "dimensional stability." Structures in space also need to be able to withstand its uniquely harsh environment. This is considered a material's "environment stability." In space, this means that the material can remain stable despite the presence of radiation and the vacuum of space. The most important properties of a material to be used in space are strength and rigidity. While in orbit, space objects are subjected to forces massive enough to tear apart weaker structures. The launch alone can put a material under forces that are as much as three times the force of gravity. Upon reaching space orbit, the satellite must retain operations in a microgravity environment. Structures with human habitation, such as the International Space Station, must also withstand the cabin pressure of oxygen, which can exert a force of up to fifteen pounds per square inch on a surface. Such a huge variation in gravitational exposure requires materials with a rare basket of qualities. Even a minor leak can lead to a domino effect of fracturing and catastrophic failure. There is also plenty of space junk in orbit, which effectively act as high-speed projectiles and pose additional hazards to structures in space. Meteorites and other natural objects can also collide with space infrastructure.

Aluminum alloys are the best-suited metals to withstand this range of stresses while also being light enough to be launched into space at minimal expense. Keeping in mind that even with the massive reduction in payload cost with the new SpaceX Falcon 9 rockets, it still costs at least $2,700 per kilogram to send material into orbit (when the Space Shuttle was in operation, it could launch a payload of 27,500 kilograms for $1.5 billion, or $54,500 per kilogram).[9] However, spaceflight has also benefited from the wondrous properties of a nonmetallic material that had its origins in my home state of Delaware. While working at the chemical giant DuPont in 1964, Stephanie Kwolek and her team developed a novel lightweight fiber with unparalleled properties of strength. The official name of this material is polyparaphenylene terephthalamide (K29), but we know it commonly as Kevlar. Although popularized for use in bulletproof vests, Kevlar is a fantastic aviation material and has given metals a real run for their money. Originally developed to strengthen tires, Kevlar has grown to be one of the company's bestselling products. Although she did not accrue direct financial benefits from the material's global sales, Kwolek's revolutionary "antimetal" invention got her inducted into the National Inventors Hall of Fame, and she was awarded the National Medal of Technology and Innovation.

Exploration of Earth's "inner space," the oceans, also requires the same kind of tough material properties as exploration of outer space, and here too the aluminum industry played a central role. Here it was Alcoa's main competitor, the Reynolds Metal Company, that took the lead. The company had its origins in the tobacco fortune of the cigarette tycoon R. J. Reynolds. His nephew, the enterprising young lawyer Richard S. Reynolds (R. S. Reynolds), worked at the company for some years and developed an all-metal, moisture-preserving container called the Prince Albert Tin, which became a major product-delivery improvement for the industry. Having whetted his appetite for the metals business with this invention, R. S. left R. J. Reynolds Tobacco in 1913 to start the Reynolds Corporation with his two brothers. Initially, they made household cleaners until the start of World War I, when soap was deemed an "unessential" luxury. When the war ended, the Reynolds brothers sold the Reynolds Corporation, and R. S. turned his attention back to the creation of a new foil-packaging company. In 1919, R. S. Reynolds founded the United States Foil Company, which later became the Reynolds Metals Company, to manufacture tin foil for cigarette packaging. When the price of aluminum came down thanks to the Hall-Héroult process, Reynolds replaced tin with aluminum.

Before the start of World War II, R. S. Reynolds contacted President Franklin D. Roosevelt regarding the need for a greater national aluminum supply. He was asked to testify before the U.S. Senate committee investigating the defense production effort. Soon thereafter, R. S. Reynolds leveraged much of his assets for a loan from the government to engage in further production, and by the time the war ended, Reynolds Metals had become the second producer of basic aluminum in the United States. During this wartime spurt of innovation, John "Louis" Reynolds, one of the founder's sons, became keenly interested in using aluminum alloys for submarine production. Ironically, he was the executive director of the foil division, which produced the most delicate and crunchable of aluminum products—hardly the ideal material for a submarine, which would need to withstand the implosive pressures of the oceans' depths. In an interview given to *Time* in 1943, Louis Reynolds and his brothers shared their interest in finding innovative uses for aluminum scrap. Behind the scenes they began to work with General Dynamics to develop a submarine that was primarily made from aluminum. Returning to the surface would be much easier for such a lighter vessel. Aluminum's strength-to-weight ratio exceeds that of steel, so a 6.5-inch-thick (170 mm) shell could withstand pressures of 7,500 pounds per square inch (52 MPa) at around 17,000 feet below sea level.

More than two decades later, in 1964, *Aluminaut* was released to the world as the first submarine vessel to be made from aluminum. Although it was labeled as an experimental vessel, the eighty-ton, 15.5-meter (51 feet) crewed deep-ocean research boat was fully operational and undertook some historic missions. It was operated from 1964 to 1970 by Reynolds Submarine Services, doing contract work for the American Navy and a range of research organizations, including the Woods Hole Oceanographic Institution in Massachusetts and Jacque Cousteau's marine biology research enterprise. *Aluminaut* is perhaps best known for helping recover a lost unarmed U.S. hydrogen bomb in 1966. During an Air Force collision over Palomares, Spain, a 1.45-megaton-of-TNT-equivalent thermonuclear bomb (Teller-Ulam design) was lost in the Mediterranean Sea. Seven crew members were killed in the midair crash of a B-52 bomber and a KC-135 refueling plane. The crash dropped three thermonuclear bombs on land and one in the sea. *Aluminaut* was sent to recover the seaborne bomb. The vessel is still functional and is regularly serviced in its retirement home at the Science Museum of Virginia in Richmond—which was for years the city that hosted the world headquarters of the Reynolds Metal Company. However, the rise of Reynolds was only possible because of the direct regulatory challenge posed to the dominance of Alcoa—the industry pioneer in the aluminum sector during the wartime years.

MONOPOLY POWER AND THE MILITARY-INDUSTRIAL COMPLEX

The world wars spawned the "military-industrial complex," which even a military general who became president—General Eisenhower—warned us about in his final speech. This speech was prescient at many levels regarding the influence of primary industries in charting the course of development options for a postwar era. Some of the key features of relevance to the use of metals like aluminum in the vast defense establishment are as follows:

> American makers of plowshares could, with time and as required, make swords as well. But now we can no longer risk emergency improvisation of national defense; we have been compelled to create a permanent armaments industry of vast proportions. . . . In the councils of government, we must guard against the acquisition of unwarranted influence, whether sought or unsought, by the military-industrial complex. . . . Today, the solitary inventor, tinkering in his shop, has been overshadowed by task forces of scientists in laboratories and testing fields. In the same fashion, the free university, historically the fountainhead of free ideas and scientific discovery, has experienced a revolution in the conduct

of research. . . . Yet, in holding scientific research and discovery in respect, as we should, we must also be alert to the equal and opposite danger that public policy could itself become the captive of a scientific-technological elite. . . . As we peer into society's future, we—you and I, and our government—must avoid the impulse to live only for today, plundering, for our own ease and convenience, the precious resources of tomorrow. We cannot mortgage the material assets of our grandchildren without risking the loss also of their political and spiritual heritage. We want democracy to survive for all generations to come, not to become the insolvent phantom of tomorrow.[10]

Large-scale sourcing of any material for a single customer can lead to a concentration of power and entrench established corporations, which are able to provide such supply in times of urgent need. Alcoa benefited from such a confluence of circumstances during World War I and continued to do so during the interwar period. As a bastion of free-market capitalism, to prevent such monopoly power the United States has developed regulatory systems that date back to the Gilded Age. Many nineteenth-century billionaire tycoons owed their wealth accumulation to material monopolies over key resources: oil in the case of the Rockefellers, lead mining in the case of the Guggenheims, and steel in the case of Andrew Carnegie. Many industrialists of this era would form a panoply of "trusts" that were interconnected and operated as a stealth mechanism for wielding monopoly power. Senator John Sherman of Ohio, who had also served as secretary of the Treasury and secretary of state in the post–Civil War period, became a champion of breaking up such trusts, which could mitigate competition and fix prices to the detriment of consumers. With an overwhelming bipartisan majority (only one dissenting vote), the Sherman Anti-Trust Act was passed in 1890.

It was under this law that the federal government began to take note of the immense power that Alcoa wielded as a supplier of aluminum during World War I. The company was the only producer of virgin aluminum ingots in the United States, pursuant to two patents. With rising demand, Alcoa increased its production of virgin ingots to keep up with the demand. In 1928, the company also spun off a Canadian subsidiary called Aluminum Limited to take over Alcoa properties outside the United States. Limited was part of the Alliance, a Swiss corporation created through an agreement between companies from France, Germany, and the United Kingdom. The Alliance members entered into agreements in 1931 and 1936 concerning aluminum manufacture and sale and set production quotas for each member that included aluminum imports into the United States.

Protesting such arrangements at an international level, the U.S. government under President Franklin D. Roosevelt sued Alcoa and Limited for violating the Sherman Act. The trial began in 1939, just as military demand for aluminum was also ramping up for World War II. The government alleged that Alcoa was monopolizing interstate and foreign commerce with respect to the manufacture and sale of aluminum ingots and that Alcoa and Limited had illegally conspired to form a monopoly and restrain domestic and foreign commerce regarding the manufacture and sale of aluminum ingots. The district court dismissed the complaint by calculating Alcoa's market share at 33 percent, by including secondary ingots (recycled scrap, which was often handled by other suppliers) in its market definition. The district court also found that Alliance's 1936 agreement did not materially affect U.S. commerce or trade because the amount of ingots imported in 1936 and 1937 was higher than in previous years. The appeal from the U.S. government went to the U.S. Supreme Court, but in a bizarre attendance technicality, the court was unable to attain a quorum of six justices required to hear the case, and it went back to the Second Circuit Court of Appeals. When the case was eventually heard in 1945 by the Circuit Court, in a shorthand summary of the law Judge Billings Learned Hand issued the verdict "that percentage (90%) is enough to constitute a monopoly; it is doubtful whether sixty or sixty-four percent would be enough; and certainly thirty-three per cent is not."[11] The government had leveled around 140 charges against Alcoa of monopolizing sixteen markets. In addition to the horizontal control of production, the case also attacked the company's vertical control of bauxite mines, energy and water infrastructure, and alumina refining capacity.

A vilification of the company at an unprecedented scale followed. The *New Yorker*'s "Reporter at Large" column claimed Alcoa featured "all the Lombroso stigmata of an octopus," referring to the dubious work of the Italian criminologist who suggested that there were inherent criminals with key attributes of form and behavior. Such stigmatization of industrial corporate power is often linked to a resentment of specific personalities who accumulate immense wealth because of the structural features of the monopoly. In the case of Alcoa, the Mellon financiers became the focal point of scorn, similar to what Bill Gates, Elon Musk, or Jeff Bezos are the target of in this day and age. The situation was further complicated by the fact that Andrew Mellon had served in the Republican governments of Harding, Coolidge, and Hoover as secretary of the Treasury from 1921 to 1932. This was a rollercoaster period of American history, from the financial giddiness of the Roaring Twenties to the start of the Great Depression. Alcoa became a particular political target for the Democratic administration of President Franklin Roosevelt, who referred to Andrew Mellon as "the master-mind behind the malefactors of great wealth."[12] He borrowed the term "malefactors of great wealth" from his presidential uncle Teddy Roosevelt—a Republican—who had referred to predatory capitalism with this term at a speech in Provincetown in 1907.[13] The ways that industrial might can create such asymmetries of power and concomitant concerns of political influence remain to this day a challenge for regulators. This case is still considered a landmark antitrust case, and generations of lawyers have written case commentaries on the verdict.[14]

Critics of the court's decision have argued that before World War II Alcoa had kept the price of primary aluminum at a level compatible with the maximum expansion of a new market, which was in its own self-interest as well. Profitability during this period required efficiency and productivity improvements and, hence, innovation. The company was thus a sort of "benevolent monopoly" rather than a "coercive monopoly," as it could not set price and production policies independent of potential market-competition forces. Indeed, only because the company focused on efficiency and cost reductions rather than raising prices was it able to maintain its competitive position as sole producer of primary aluminum. If Alcoa had tried to exercise monopoly influence to increase profits through price increases, it could have possibly found itself making its own new competitors. However, the scale of investment, particularly in the energy infrastructure needed for aluminum production, does diminish the possibility of too many new entrants. In an anthology published by the iconic free-market libertarian Ayn Rand, a notable retrospective note of sympathy for Alcoa was offered by the long-time chair of the U.S. Federal Reserve, Alan Greenspan, as follows:

> ALCOA is being condemned for being too successful, too efficient, and too good a competitor. Whatever damage the antitrust laws may have done to our economy, whatever distortions of the

structure of the nation's capital they may have created, these are
less disastrous than the fact that the effective purpose, the hidden intent, and the actual practice of the antitrust laws in the
United States have led to the condemnation of the productive and
efficient members of our society because they are productive and
efficient.[15]

The court verdict was a major blow to Alcoa, but the timing
of the decision occurred at a time when the war economy was
winding down, leaving a surplus of aluminum on the market.
The government had to set up a postwar Surplus Property Board,
which was tasked by Congress to dispose of around $90 billion
dollars' worth of excess war property in 1945 (equivalent to around
$1.4 trillion dollars today). There was plenty of metal in this col-
lection of wartime miscellany. The court's antitrust ruling led
Alcoa to divestments that were not as consequential as they
might have been in other times. The company adapted far better
than what many would have expected thanks to the astute management of its corporate executives. The legacy of Arthur Vining Davis at Alcoa, in the United States, and his brother Edward
Davis, who became the head of Alcan in Canada, was remarkable in its influence across future generations of leadership at the
company in its binational form.

Among Alcoa's notable twentieth-century leaders was Paul
O'Neill, who later became the seventy-second U.S. secretary of
the Treasury under the first term of President George W. Bush.
Coming from a military family, O'Neill started his career in public service with the Office of Veteran Affairs and the Office of
Management and Budget, where he also served as deputy director from 1974 through 1977. His corporate career began soon after
his OMB job; he took a series of executive jobs with International Paper Company, eventually becoming its CEO from 1984

to 1987. International Paper grew forests as a renewable resource and made paper from their wood. This may seem to be a very different sort of business than Alcoa, which mined a metallic resource that could not be "grown" or "cultivated" on land by humans. However, both International Paper and Alcoa were vertically integrated and involved attempts to integrate recycling as a means of charting a more sustainable trajectory. O'Neill was thus tapped to lead Alcoa at a time when the company was facing major concerns about its safety record and the treatment of its workers. He also showed the industry that it could do well by doing good and by having an uncompromising stance on prioritizing safety. During his tenure as CEO, Alcoa's market value increased from $3 billion in 1986 to $27.53 billion in 2000, while net income increased from $200 million to $1.484 billion.

As an example of his manifest attempt to change the culture of heavy industry around health and safety, O'Neill would personally get involved in occupational accident investigations on site. In one such episode, when an employee was killed as he jumped over a safety barrier to repair a factory machine, O'Neill called an emergency meeting so he could personally investigate what happened. Within seven days, safety rails at all Alcoa facilities were painted a bright yellow, and workers and managers received additional training about how to handle needed repairs to equipment. As Ronald Humphreys notes:

> Paul knew that accidents happened when the manufacturing process went wrong—something bad for both workers and Alcoa. Whenever an accident happened, Paul encouraged workers and quality control experts to work together to figure out what went wrong. It turned out that worn out and poorly maintained equipment was a major cause of both accidents and poor-quality aluminum. Workers were empowered to report equipment that

needed replacing, and as a result, accidents declined dramatically and profits soared.[16]

Despite continuing antitrust lawsuits in subsequent years, Alcoa and Alcan both expanded globally and built remarkable bridges of industrial diplomacy in the postwar period between erstwhile enemies. In his memoirs, titled *Expect Miracles*, David Culver, the renowned chief executive of Alcan, recounts how the company's executives were like "Jesuits" taking the message of aluminum to the world. As with any theological analogy, there was the danger of taking evangelism to the point of expropriating culture, yet such spirited devotion to a "cause" also spurred genuine human exchanges in distant lands that would otherwise might not have found common interests in connecting. The geological determinism of ores and energy availability made necessary such exploration. Market forces and technical prowess also motivated multinational businesses to reach far and wide. When relations between the West and Japan eased, executives from metal companies were among the first at the scene with contract opportunities. Indeed, some of the motivation for a postwar rapprochement between the United States and Japan could be tied to the material-manufacturing supply nexus of the automotive and electronics industries.

Aluminum was a key component of the automotive sector, and the metals industry was quite keen to sell their product to Japanese manufacturers. Commercial cars and aircraft played an important role in moving aluminum away from its reliance on the military-industrial complex, in so doing raising aluminum's profile as an agent for peace. The aluminum industry's role in forging better relations between the West and the East—particularly, Japan, China, and Russia—was significant. However, the most enduring legacy aluminum had on the political economy of war and peace was in developing countries' attempts to emerge from colonialism.

MINERAL NATIONALISM AND ECONOMIC DEVELOPMENT

Who owns planetary resources? This is a question central to war and peace in human civilization since time immemorial. Access to water, primary forests, and minerals have all forced fundamental philosophical and political conversations about morality and our notions of "rights." While every culture has formulated codes to regulate collective behavior, the Romans were perhaps the best at packaging and marketing their doctrines. In the second century CE, the Roman jurist Gaius put together what is believed to be the most comprehensive set of legal principles. These Institutes, as they were called, were later appropriated by the Emperor Justinian in the sixth century and are often referred to by his name. Justinian's contribution was to add statehood to these legal principles by asserting a *ius gentium*, or law of nations, based on what he called "the source of morality and the true foundation of all civic laws," which contained "the idea of all mankind as forming one natural community of which all are citizens."[17] The idea of having a "natural universal law"—which many indigenous hunter-gatherer societies have claimed for millennia—was motivated in the West by the European Enlightenment, when philosophers sought to mitigate the authority of the church. More recently, this concept has also been used to promote international law by the United Nations. However, such efforts have always been in tension with the human need to assert individual "ownership" and territory. Territoriality has more visceral roots, perhaps stemming from our animal instincts for survival.

The Codes of Justinian in Roman law established common property of citizens, including the "air, running water, the sea, and consequently the shore of the sea," and distinguished between *res publica* (state property), *res privatae* (private property), and two categories of common property: *res communes* (open access) and *res universitatis* (community property). They also had the category of *res nullius*, the property of no one, which could potentially be claimed by others under certain circumstances.

As minerals were essential ingredients of state power in terms of weapons production and thus the protection of citizens, Roman law granted mineral rights to the state. Much later, British Common Law was predicated under the doctrine of *Cuius est solum, eius est usque ad coelum et ad infernos* (Latin: "whoever's is the soil, it is theirs all the way to Heaven and all the way to hell"). The Roman glossator Franciscus Accursius had also noted this *ad coelum* doctrine in the thirteenth century. This idea gained more traction as the British adopted the principles of private entrepreneurship and colonized America, but in the homeland itself, a different concept, that of the "Commonwealth," took precedence. Under that idea, you could own the soil and surface of land but not have rights to the minerals, because of their primacy for the common good. The extraction of these minerals could therefore also hold precedence over surface land rights. This "apportionment" of property rights has to this day been one of the primary mechanisms by which the colonial enterprise and its postcolonial developmentalist doctrines have come into massive conflict, the argument being that minerals are a primary resource that the state can use for the development of the entire nation, and therefore the state should hold dominant property rights over them as a means of both material usage for infrastructure and defense but also for tax revenue.

The story of aluminum's extraction from bauxite exemplifies these tensions in a way more definitively than other minerals do because of the high energy needs of processing and the propensity for vertical integration within the industry. However, as noted by the authors of an important anthology on the political economy of aluminum:

> The vertical integration in the sector is also asymmetrical in the sense that there is substantial differentiation between segments of the value chain. Upstream in the value chain, the industry is global and highly integrated, but downstream, beyond the smelting stage, the aluminum companies are really engaged in multi-domestic businesses because of the diversity of the local markets, high transport costs, strong competition from independent fabricators, and the requirement for intensive customer contact.

Furthermore, bauxite has what resource economists would call "asset specificity"—each location of ore has geochemical characteristics that require specialized processing, and refineries need to be calibrated accordingly, which can be expensive to do.

France and North America dominated the aluminum production in the early days of the metal's industrial ascendance. However, with rapid rise in demand, mineral exploration for bauxite ore in distant lands began to take priority. The northeastern region of South America's ancient tepui mesas and deep valleys had the perfect conditions for the weathering of laterites to concentrate bauxite deposits. The indigenous inhabitants of this lush region replete with rivers called it Guyana—meaning "land of water." In a visit to the region in 2019, I had a chance to take a low-altitude flight across the vast forests of the Guyanas. The only punctuations in the foliage one finds are mining sites, either riparian gold-mining sites being extracted by artisanal or small-scale miners or larger bauxite mining pits.

Apart from its minerals, the region also had immense strategic value for European powers as a gateway to the Americas, and thus several set up colonial outposts there. The French were the first to find bauxite ore in their Guyanese colony southwest of Cayenne, as early as 1870. Soon thereafter, the British and the Dutch also found bauxite deposits in their respective Guyana territories. Alcoa first mined bauxite in Dutch Guyana—in what is now the country of Suriname—in 1917 and began producing alumina there in 1941. The same year, the United States made an agreement with the Dutch government to send two thousand American troops to the territory to protect its bauxite deposits. German U-boats and Italian submarines were sent to the Caribbean to disrupt Allied supplies of oil and aluminum as early as 1939. In 1942, a more aggressive Caribbean campaign was launched by the Germans under the leadership of Grand Admiral Karl Donitz to focus on oil and aluminum supplies. Oil refineries in the Dutch colonies of Aruba and Curacao were targeted first, as they supplied over half a million barrels of oil daily to the Allies.

Saboteurs were transported by submarine to target two cryolite plants in the United States as well. There were also reports of Nazi activity on the Brazilian border with Dutch Guyana. President Roosevelt was conscious of this challenge and arranged a special arrangement with Queen Wilhelmina of the Netherlands to undertake joint military protection of the bauxite mines in the colony, from which more than 65 percent of the American bauxite supply was sourced. The aluminum demand surge spurred by World War II brought the bauxite of the Guyanas into high demand but also created a political economy of extraction that often shortchanged local populations. Suriname remained the largest global producer of bauxite up until the 1960s. The other major producers in the Caribbean region were British Guiana, where bauxite mines supported Alcan's Canadian smelters, which also served U.S. war production. As the Guyanas gained independence, the question of how mineral investments would be managed by the new nation-states reactivated the debates over state ownership of mineral rights versus the private investors, who had predated independence. British Guyana gained independence in 1966 and Dutch Guyana in 1975.

Alcan had been the major investor in British Guyana through its subsidiary Demba, which had mined bauxite along the Demerara River since 1917 (when it was still a part of Alcoa). After independence, Guyanese politics had to contend with the friction between ethnonationalist and resource nationalist factions. Guyana had a diverse population, given its history of slave trade from Africa, indentured labor from India, and small indigenous communities. Mining was particularly opposed by the Maroon communities—descendants of African slaves who escaped into the forests and often intermingled with indigenous communities, developing their own distinct cultural identity. Guyanese politics divided along ethnic lines, and the modalities by which resource extraction should be allowed were fundamental points of contention. The Guyanese government claimed in 1970 that over its fifty years of operation, Alcan had mined close to $1 billion worth of bauxite, and Guyana had received only 1.3 percent of that amount. At the same time, Demba claimed that since it began operations in 1919, it had paid $116 million in royalties, taxes, and other payments to Guyana's governing authorities, or roughly 13 percent of its proceeds from the sale of bauxite. It had also spent $208 million on wages and $175 million in manufacturing infrastructure. It claimed that only 9 percent of its income remained as profit. Core to these conflicting accounting statistics was the issue of the nationalization of mineral investments. Alcan also sought assistance from the Canadian government in preventing such an outcome in Guyana. Unlike the American government, which would often mix business investment with direct development assistance or other forms of leverage, the Canadians were much shyer about proffering a quid pro quo. Eventually, Guyana's bauxite mines would see a variety of foreign owners in partnership with the state. The opportunity to establish large metal production capacity has yet to be realized, but with the development of major offshore oil and gas reserves as well as hydropower potential, the country's young president Irfan Ali claimed in 2022 that greater potential for added value could soon be realized.[18]

The story of bauxite in the Caribbean started with the Guyanas but becomes far more complex in Jamaica, which has a much higher population density and where the mines competed for land use with farmers. Although many of the geopolitical interventions between companies and corporate interests from the Americas were similar to what we saw in Guyana, the level of domestic opposition was far greater because of land-use and property rights issues. The discussion of Jamaican bauxite merits this context of land resource impacts and potential for restoration processes and will be discussed in the final chapter.

Across the Atlantic, dramas of mineral politics also played out in the French colonial region of African Guinea, which was emerging from a period of colonialism and contained some of the world's highest-grade bauxite. During the Cold War, Guinea was courted by Russia in this crucial period of state building in the 1960s, and even now the Kindia Bauxite Company in Guinea is the largest foreign asset of the Russian state-owned aluminum giant Rusal. In 2018, when President Alpha Condé suggested a desire to change Guinea's constitution to give himself a third term as president, the Russian ambassador to Guinea backed a

change of the constitution to allow the octogenarian president
to "reinvigorate" the country. Conde had received Russia's Order
of Friendship, and the military coup that ousted him in 2021
unsettled the status quo. The junta controlling the country has
been resolute in supporting mining investment but has also stated
that one of its key reasons for intervention was a perceived loot-
ing of rents from the country's vast mineral wealth by foreign
companies and corrupt elites. Bauxite forms a part of the political fabric of Guinea and has been both a blessing and a bane for
the country's fortunes.[19]

Efforts at overcoming the proverbial "resource curse" had
ostensibly been the mainstay of the Conte government's court-
ing of foreign donors. Just before the coup, the country had been
validated by the Extractive Industries Transparency Initiative
(EITI)—an international mechanism to certify revenue transparency of countries, largely supported by the Norwegian government. In an interesting twist to conventional wisdom about
the better transparency of democracies, the EITI board, chaired
by the former UNDP head and former New Zealand prime
minister Helen Clark, gave Guinea a high score for compliance
with standards in its February 2022 board meeting, despite the
military coup.[20] As Africa's largest bauxite producer, Guinea
is heavily dependent on its extractive sector, which accounts
for around 30 percent of government receipts and for nearly
78 percent of total export earnings. Bauxite comprises more than
60 percent of these revenues.

During a field visit to Guinea in 2022, I met with a variety of
stakeholders who have been trying to move the needle from
transparency to impactful development. The EITI's high score
for Guinea had been a first step, but the abject poverty in the
country and the lack of infrastructure remains alarming for visitors. Just reaching the aluminum-mining region of Boke
involves traversing roads that are corrugated and crushed with

erosional fatigue. Many expatriates from this region who now
have jobs in Conakry or are overseas have joined forces to try to
take on small development projects in the bauxite region like tree
planting or plastics collection for recycling.

An organization called Kakande Kansinbak has emerged
from the "children of Boke." I met with one of the founders of
the organization, Paul Dourabangoura, in Conakry. He grew up
in Boke and is now a finance professional who managed to collect around $12,000 per year from donations; locals also volun-
teer their time. Companies could have been held to further
account for developing infrastructure, and the public sector could
have invested more from the tax income that bauxite provided.
A former mining minister of Guinea with a deep understanding
of the sector noted to me that the challenge of bauxite was that
its value chain was very capital intensive and developing downstream capacity is highly energy intensive.

While Guinea had bauxite, nearby Ghana was endowed with
massive hydropower resources. The development of the massive
Volta River hydroelectric scheme to power an aluminum smelter
in what was then the British Gold Coast Colony created the
world's largest artificial lake, displacing over eight thousand people. As the decolonization of Ghana from Britain proceeded,
the American aluminum giants Alcoa and Kaiser joined forces to
finance the Akosombo dam, which would provide hydroelec-
tricity to this smelter. The project had its origins in the British
Colonial Development and Welfare Act of 1940. The postcolonial
leadership also realized, however, that infrastructure development would be needed for the country to improve its citizens'
quality of life. This was the Age of Dams—where the term "conservation" had a very different connotation than it does today in
our environmental lexicon. Conserving water meant to dam and
contain rather than "waste" the freshwater by letting it flow into
the sea. The Volta Aluminum Company was thus conceived with

this resource-exploiting perspective of water and materials in mind.

Eventually, in this case resource nationalism would prevail, with the Volta Aluminum Company becoming a state enterprise in 2008. However, for over forty years, the investment from Kaiser Aluminum created what the historian Stephan Meischler has called "an American island" in West Africa.[21] The dam and its aluminum enterprise were transformative for the newly independent state. While the dam also provided electricity for Ghanaians, it was largely motivated by the desire for the country to showcase its industrial might. Not only could it show the world that it mined gold (for which it was well known) but that it could also process and refine metals in a "value added" process. The dam's provision of hydropower to industry was also intended to diversify the economy beyond its dependence on cocoa. The prized forest pods of the cacao tree that flourished in the Ghanaian soil had produced many Cadburys chocolate bars for the British Empire but not delivered meaningful development for the colony. However, aluminum production did not lift Ghana out of poverty, as many hoped it would. Just as cocoa turned into chocolate for Western consumers, Ghana's aluminum went to Western markets. Ghana remained largely peaceful, in comparison with other West African states like Sierra Leone, Liberia, or Cote d'Ivoire, but peace and development did not necessarily go hand in hand. Meanwhile, the rise of worldwide aluminum demand continued. The question that many began to ask was whether this rising demand for aluminum materials was being subsidized by the low labor costs and environmental damage of extraction in distant lands.

6

ALUMINUM FOR ALL

The Invention of a Household Metal

*As well might it be said that because we are ignorant of the laws
by which metals are produced and trees developed, we cannot
know anything of the origin of steamships and railways.*
—Alfred Russell Wallace

I n the classic 1986 science-fiction feature film *Star Trek IV:
The Voyage Home*, the crew of the USS *Enterprise* are confronted by an enigmatic probe that is draining energy
resources from Earth as it tries to communicate using whalesong:
a futile task, because in the twenty-fourth century, these creatures are extinct. The crew must travel back in time to find a
humpback whale and transport it forward to the twenty-
fourth century to reply to the distress call of the probe. At this
point, "transparent aluminum" enters the plot as the material that
could be used to build the tank to transport the whale. This is
the same material of the future that the windows of the Enterprise are made of. Within the Star Trek saga, *The Voyage Home*
was the first film with a distinct environmental message—
far-fetched as the plot may be. An extinct species helps save the
planet from ruin, but this anachronistic encounter is mediated
by a unique material that is a product of human innovation.

One may ask why the writers of *Star Trek* chose aluminum as the metal of choice for their rendering of this fictional material. The vast array of products that aluminum has spawned most likely played a role, as well as the perception of the metal as thin, light, and flexible. Although the free-floating electrons in metals generally prevent photons of light from passing through (a necessary condition of transparency), compounds of aluminum, particularly alumina (an oxide of the metal), can show such properties. Synthetic sapphires used in certain electronic screens are essentially a form of transparent alumina. There are now also a host of other aluminum compounds that have some of the properties of the ostensibly fictional material. In 2009, some scientists at Oxford even created a transparent form of metallic aluminum for a few seconds by bombarding the metal with the world's most powerful soft X-ray laser; they declared it to be a "new form of matter."[1]

Often aluminum compounds exist in various forms in combination with calcium and silicon compounds, and their variety can be shown in what is generically called a "ternary phase diagram" (figure 6.1). Aluminum oxide, Al_2O_3, occurs naturally in the form of corundum, the crystalline phase of Al_2O_3, which exists as natural rubies and sapphires, depending on the color given to the crystal by trace elemental impurities. I was first alerted to the salience of this mineralogical trinity while visiting the campus of Pennsylvania State University and walking into the Deike Building, which houses the College of Earth and Mineral Sciences. Its lobby holds a stark black slate mural that covers a massive wall and lays out the story of human interactions with minerals.

Etched in one corner of the mural is this ternary phase diagram, laser-crafted by the artist Mick Fleck, who chose to show how research could harness an assortment of ceramic materials

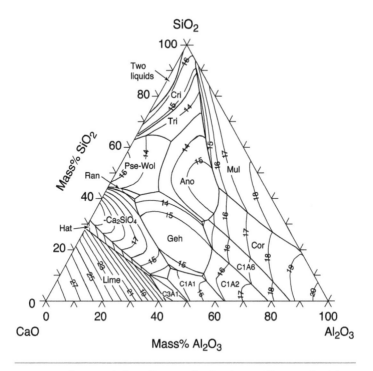

FIGURE 6.1 A mineral phase diagram for aluminum, silicon, and calcium minerals like the one etched into the wall of the Minerals Museum at the Pennsylvania State University. Numbers indicate temperature in degrees Celsius.

Source: Figure adapted from established geological chart.

from these simple minerals. Aluminum, even in its pure earthen form, has given us a vast range of products. Human ingenuity in its primary and primordial form relates to how we can take such natural assemblages of minerals and use them for a variety of new products and materials. Aluminum's versatility in this regard is exemplary, thanks to its physical properties. However, although the phase diagram shows the range of material "recipes" emerging from Earth's natural cauldron of three key minerals, aluminum metal in its refined form also requires fine tuning with other ingredients for maximum product versatility.

How aluminum's range of products gained traction is an exciting story that dates to the early days of its discovery. Soon after Napoleon III patronized Paul Héroult's aluminum production, Wilhelm II, the king of Prussia and emperor of Germany, commissioned a research center at Neubabelsberg, south of Berlin, to research and develop alloys of the metal. The center started operations in 1902 and employed an erudite metallurgist named Alfred Wilm (1869–1937). The work of Wilm in developing some of the early alloys of aluminum was so remarkable in its later potential for product development that the management historian Quentin Skrabec noted in his history of aluminum in America that "without Wilm, aluminum would have been limited to jewelry and steel production."

Wilm's key innovation was to develop an alloy called duralumin—derived from the phrase "durable aluminum." He discovered that the process of "quenching" molten aluminum with a range of metals in water can lead to material properties far more versatile than those derived from alloy formation in combined smelting. To this day, the range of duralumin alloys contain small amounts of copper, manganese, and magnesium, which are the mainstay of most aluminum compounds. Wilm is a relatively forgotten figure in the annals of aluminum and did not receive any patent royalties from his work. In 1919, he left his metallurgical work and became a farmer to focus his energies on soil rather than foil. But there were many other entrepreneurs to take up his mantle, bring duralumin to market, and liberate its potential for product diversity. In 1894, a magazine called *Aluminum World* emerged with support from the nascent industry, and in its millennial-edition issue at the end of 1900, it published the following ten uses for the metal (the "etc." is from the original publication list):

1. Manufacture of utensils and in the arts, as kitchen utensils, tableware, bottles, watch hands, trimmings for books, wagon frames etc.

2. Articles for military use, as equipment for soldiers and general warfare implements

3. Articles for marine and aeronautical purposes

4. Instruments and apparatus for the surgical purposes and sick room supplies, such as respirators, syringes, catheters, coffins, artificial teeth plates etc.

5. Instruments for mathematical, physical, optical and chemical uses.

6. In the wire industry for wire brushes, baskets hooks and eyes, eggbeaters, dog muzzles, bird cages, etc.

7. For foils, bottle caps, etc.

8. Aluminum powder and paint

9. As a reducing agent

10. For lithographic printing.[2]

Realizing this full potential of product development for aluminum involved some competition between aluminum companies. Inventors are not always innovators beyond their very particular modes of interest. This was true to some degree for Charles Hall and the early entrepreneurs who invented the refining process for aluminum. Their intellect was focused to a large degree on the production process. The vast range of aluminum products that now animate our lives came about with a fresh set of entrants to the arena.

OLIGOPOLY INNOVATION

The advent of aluminum alloys was important for Alcoa, but with the war consuming most of the supply, there was scant incentive to fully realize the potential profusion of products from this metal. The settlement of the monopoly lawsuits empowered other industrialists to further their momentum in product development to compete with Alcoa. In this regard, R. S. Reynolds was far more of a pioneer than the Davis brothers at Alcoa and Alcan. Recognizing the strength and malleability of aluminum, Reynolds focused his product innovations on the American kitchen. Apart from the panoply of utensils and cutlery, his revolutionary development of Reynolds foil in 1947 created a product that brought aluminum into almost every American household. Its ubiquity accustomed consumers to using the metal, which would subsequently find its way into our lives through beverage cans and food containers as well. The original idea for the wrap for a variety of food products may have originated in Europe. Developed as a replacement for tin foil, the Swiss engineer Robert Victor Neher took out a patent in 1910 for the continuous rolling process and opened the first aluminum rolling plant in Kreuzlingen, Switzerland, and by 1911, the Bern-based Tobler firm began wrapping its chocolate, including its unique triangular chocolate bar, Toblerone, in alufoil. By 1912, alufoil was being used by sister Swiss company Maggi to pack soups and stock cubes as well.

The European market for consumer products at the time was fairly removed from the U.S. market, and Reynolds became a strong competitor of Alcoa. In 1940, Reynolds Metals began mining aluminum ore at the aptly named town of Bauxite, Arkansas, which had a history of mining dating back to 1890, when the first ore deposits were discovered. The company also opened its first aluminum plant near Sheffield, Alabama, in 1941. With the success of antitrust litigation against Alcoa, Reynolds Metals Company leased and later bought six government defense plants that were up for disposal. In 1948, R. S. Reynolds turned the leadership of Reynolds Metals over to his four sons but retained a seat as the chairman of the company until his death in 1955, when he was remembered in obituaries as the "Aladdin of Aluminum" and the "Poet Industrialist." With growing wealth, the company designed an iconic headquarters on the outskirts of Richmond, Virginia. The modernist building, designed by the architect Gordon Bunshaft, contained 1,235,800 pounds (560,500 kg) of aluminum, 400,000 pounds (180,000 kg) in the exterior cladding. This grandiose investment competed with Alcoa's headquarters in downtown Pittsburgh, which had been built five years earlier and sported a massive amount of aluminum siding.

David P. Reynolds, the son of R. S. Reynolds who served as the last family member to lead the aluminum company, earned a reputation as an environmentalist, promoting the reuse of aluminum as a solution to litter and waste. In 1987, he received an award from Keep America Beautiful for pioneering efforts in recycling. He firmly believed that regardless of the need to diversify suppliers of aluminum, the metal itself should monopolize human material consumption across a wide swath of products. In his obituary in 2011, the *New York Times* wrote:

> The Reynolds house was outfitted with an aluminum solar-paneled roof; an aluminum Christmas tree graced their home during the holidays; and the family freezer was stocked with foil-wrapped ice cream from the Eskimo Pie Company, a subsidiary of the family metals business. Mr. Reynolds even gave his wife aluminum jewelry, something she wore sparingly, preferring more precious metals.[3]

In an interesting twist of monopoly déjà vu, Reynolds was bought by Alcoa in 2000. In 2008, the New Zealand billionaire

Graeme Hart paid US$2.7 billion for the pacing division of Alcoa, which had its origins in Reynolds, and spun off the company, renaming it Reynolds Packaging Group, which is now headquartered in Lake Forest, Illinois. Since the purchase from Alcoa, Hart cut more than 20 percent of the workforce within Reynolds, mostly through plant shutdowns. The iconic Reynolds headquarters building is now the U.S. corporate head office for Altria—the company that emerged from the rebranding of the tobacco and consumer products giant Philipp Morris.

The third key player in the "American aluminum trinity" was Kaiser Aluminum, which had its origins in the Californian shipping business. The shipping magnate Henry Kaiser founded the company in 1946 with two reduction facilities and an aluminum rolling mill. The company subsequently expanded to include mining and manufacturing processes and the distribution of aluminum. As part of its vertical-integration model, Kaiser acquired bauxite mines and set up refineries in key locations to complement its shipping business.

Kaiser's aluminum enterprise had an ambivalent history of innovation based on synergies with its existing businesses. Given its experience of using asbestos in shipping, Kaiser also utilized the material to line vessels exposed to extreme heat during the aluminum-refining process. This was an energy-saving innovation but one with serious consequences for occupational health. Kaiser and its subsidiaries manufactured asbestos-containing products from the 1950s until 1978, which caused mesothelioma and asbestosis among their employees. To handle the financial strain caused by asbestos litigation, Kaiser filed for chapter 11 bankruptcy in 2002. When the company emerged from bankruptcy in 2006, the Kaiser Aluminum Chemical Corporation Asbestos Personal Injury Trust was established with $1.2 billion. The trust was formed to process, liquidate, and pay all valid asbestos personal-injury claims for which Kaiser has legal responsibility. In April 2019, the payment percentage was reduced from 35 percent to 25 percent, which remains substantially higher than most other asbestos trusts.[4]

While its technical innovation in aluminum production had backfired momentously, Kaiser was also involved in a social innovation. At its shipyards in Richmond, California, Kaiser, to differentiate itself from its competition, implemented the pioneering idea of offering group insurance in partnership with the medical doctor and social entrepreneur Dr. Sidney Garfield. Through initial funding from the U.S. Maritime Commission and Henry J. Kaiser's Permanente Foundation, the Kaiser Richmond Field Hospital became the prototype for what we now call "health maintenance organizations" (HMOs). By August 1944, 92.2 percent of all Richmond shipyard employees had joined the Kaiser Permanente organization, the first voluntary group plan in the country to feature group medical practice, prepayment, and substantial medical facilities on such a large scale. To this day, Kaiser Permanente is America's largest nonprofit HMO, and it all started with an attempt to satisfy the needs of an expanding workforce in the oligopolistic world of the aluminum industry.

In recent decades, Kaiser Aluminum has stopped its bauxite and alumina operations and has focused on high-grade aluminum product manufacturing, such as aluminum components for the aerospace industry. Kaiser Aluminum now owns twelve fabricating plants throughout the United States and Canada and yields more than 600 million pounds of aluminum annually. In 2020, the company reported net sales for the previous year of $1.5 billion and employed more than 2,800 people.

Oligopolies often emerge "naturally" in metal markets given the limitations of large investments beyond a few key dominant market players. Alcoa only had two major competitors, Reynolds and Kaiser, and for all practical purposes these three companies could be considered an oligopoly. In such cases, prices for the consumer are neither set by capricious monopoly power nor by the "invisible hand" of a free market. There may be some level of tacit collusion, but from a legal perspective it is far more difficult to use antitrust laws to convict an oligopoly. There may also be some benefit for the consumer in terms of balancing market efficiency with innovation of product delivery. The excess profits that come from oligopolistic collusion can be invested in research and development, which can lead to greater innovation. Whether these innovations lead to greater "stuff" for consumers or to more efficient processes for making existing products becomes a question to be curated by consumers. However, innovations, particularly when designed with emotional persuasion, can at times become tantamount to what Jessica Helfand has called "the Invention of Desire."[5] Ultimately, such desires must be constrained by planetary limits, though technology may always strive to expand those limits.

METALLIC COMBINATIONS IN THE AGE OF HIGH TECHNOLOGY

The key ecological question that critical sociologists like Mimi Sheller have confronted us with is whether the innovations that lead to a new set of products for the materials are actually a net positive for society. As she notes, "The aluminum dream so easily turns into an aluminum nightmare for those who dwell near the mines or who find their rivers dammed and valleys flooded for hydroelectric power."[6] Indeed, as we consider the range of aluminum products that arose through the postwar era and continue to do so now, the critical question of necessity continues to haunt us. However, in a world of complex supply chains with associated livelihood choices and constraints at every step, "necessity" has a more nuanced meaning than what an environmentalist's minimalism may suggest. The end product may be superfluous, but once it is tied to livelihoods, there become the "needs" of those who mine, manufacture, service, and deliver that product. I dealt with this connection between "wants" and "needs" at a broader level in one of my earlier books, *Treasures of the Earth*. Gandhi challenged us with his aphorism: "The world has enough for everyone's need, but not enough for everyone's greed." I would respectfully challenge this assertion insofar that someone's greed may help meet some other person's need. When we devise new products and services, we create economies across new supply chains, from mines and fields to markets. Whether those economies and the associated livelihoods can be replaced with other options is the key question to ask when advocating minimalism. In some cases, people may be content having less income and living off the land, but in other cases that may not be possible or preferable. In pluralistic societies, where some level of individual choice on what we consume and what we choose to be employed as may well be at odds with what is optimal for society.

The only way out of this nexus of need and greed is to find technical and social innovations and to find uses for materials that are most socially productive while also providing livelihoods. The COVID-19 pandemic forced us to think in terms of major livelihood transitions, and we realized that many transitions in employment that we hitherto did not think possible were at times even preferable. People went back for training, changed their careers, and reinvented their livelihoods with remarkable adaptability. A lot of this was possible thanks to technologies that allowed us to work from home or use rapid communication mechanisms. There is, of course, physical infrastructure, involving metals, behind all such communication networks. Finding such uses of metals, which are clearly improving our quality of life and making us a more adaptive species, is preferable to using these metals solely as trinkets and soda cans. Our efforts should be directed toward finding a portfolio of products for aluminum—or, for that matter, any material—that considers the system-wide impacts of its use and benefits and prioritizes the development of those industries.

Originally, aluminum consumer products were focused on the kitchen. It is indeed a chemical wonder how we can cook in aluminum dishes, given that the combination of iron oxides (in old stove grills and burners) with aluminum produces a spectacular thermite reaction. The protective layer of aluminum oxide that forms on the surface of aluminum is why the metal remains so eminently usable in so many different applications. The centrifuges for purifying uranium that have been scrutinized by the international community in Iran are made of aluminum for this reason, as well as for its light weight.

Since 2014, the bestselling Ford's F-150 pickup trucks use aluminum rather than sheet steel in the truck bed and cab, partly to reduce weight. The transition to a low-carbon-usage economy has made the weight of vehicles a key consideration, as fuel consumption is reduced with lighter vehicles. The aluminum industry's traction with the automotive sector also reflects consumer preference for more fuel-efficient vehicles. To meet such expectations, automakers have been using aluminum to reduce vehicle mass. The F-150's switch to aluminum for 2015 was a major component of the reduction of seven hundred pounds in vehicle weight compared to the 2014 model. However, Alcoa still faced significant challenges in selling to the auto industry because consumers were under the impression that aluminum was not as strong as steel. This was especially true of the male-dominated truck market, where the strength of steel and overall "toughness" and weight of the vehicle is part of the brand perception. Ford launched an educational campaign about the strength of aluminum, while the steel industry touted its own record of frame strength. Aluminum did not have a price advantage over steel, and automakers have had long relationships with steelmakers, for over a century gearing their production to steel components. Furthermore, steel makers are working on lighter alloys to compete with aluminum and carbon-fiber based composites, and plastics are also gaining ascendance.[7]

In addition to its structural attributes, aluminum's chemical properties are gaining further attention in unusual ways. In 2021, MIT researchers suggested a schematic for producing hydrogen using scrap aluminum and water. The U.S. Department of Energy has been studying metal-water reactions for hydrogen generation for decades, but this was the first time a prototype for a commercially viable process was developed with a circular-economy perspective in mind. Fabricated samples of pure aluminum and aluminum alloys can be treated to ensure that the surfaces of all the aluminum "grains" that make up the solid remain free of oxide deposits throughout the reaction. Reducing grain size can make aluminum powder, which can produce explosive thermite reactions, as small grains lack the impermeable oxide layer that forms and prevents the reaction. Overcoming this challenge of safety and efficacy required a remarkably innovative approach.

Next, one can "tune" the hydrogen output by starting with pure aluminum or specific alloys and by manipulating the size of the aluminum grains. Such tuning can be used to meet demands for brief bursts of hydrogen, for example, or for lower, longer-lasting flows. The work confirms that, when combined with water, aluminum can provide a high-energy-density, easily transportable, flexible source of hydrogen to serve as a carbon-free replacement for fossil fuels. To enable the hydrogen-forming reaction to occur, the researchers painted the surface of the solid with a carefully designed liquid-metal mixture. They started by combining two metals—gallium and indium—in specific proportions to create a "eutectic" mixture, that is, a mixture that would remain in liquid form at room temperature. They further refined the process through doping with both silicon and magnesium. After this treatment, aluminum can be dropped in water, and it'll generate hydrogen—with no energy input required, and the eutectic doesn't chemically react with the aluminum. At the end of the reaction, the gallium and indium can be recovered and reused, as they are relatively rare and expensive metals.[8]

Aluminum has also been the central element in an enigmatic form of matter called "quasicrystals," which led to a Nobel Prize for the Israeli physicist Dani Shechtman. The discovery of these crystals contradicted the established classical view of crystallography that fivefold—and its multiples—symmetry could not exist in extended structures. Shechtman did not dismiss this observation despite the peer pressure to do so—including from Nobel laureates like Linus Pauling. He persevered against dogma and let his empirical observations dictate his efforts. These crystals are yet another example of the unique position of aluminum on the periodic table and its potential for crafting a vast edifice of creativity. Quasicrystals exist between the amorphous solids of glasses (special forms of metals and other minerals, as well as common glass) and the precise pattern of crystals. Like crystals, quasicrystals contain an ordered structure, but the patterns are subtle and do not recur at precisely regular intervals. Rather, quasicrystals appear to be formed from two different structures assembled in a nonrepeating array, the three-dimensional equiv-alent of a tile floor made from two shapes of tile and having an orientational order but no repetition. Although when first dis-covered such structures surprised the scientific community, it now appears that quasicrystals rank among the most common structures in alloys of aluminum with such metals as iron, cobalt, or nickel. Figure 6.2 shows the structure of these crystals and the central role of aluminum.

An application involving the use of low-friction Al-Cu-Fe-Cr quasicrystals is being considered as a coating for cooking utensils, as their heat transfer and durability is better than conventional nonstick cookware. Pans with such a coating can withstand tem-peratures of 1,000°C (1,800°F) without harm. Unfortunately, the coating was sensitive to salinity and eventually would cor-rode, so the plans were withdrawn from production.[9] The

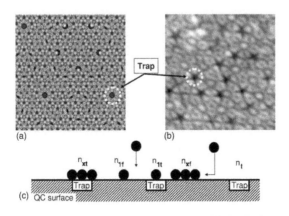

FIGURE 6.2 The fivefold surface of i-Al-Pd-Mn QC as observed through a scanning tunneling microscope, previously thought to be "impossible."

Source: From B. Unal, et al., "Nucleation and Growth of Ag Islands on Fivefold Al-Pd-Mn Quasicrystal Surfaces: Dependence of Island Density on Temperature and Flux," *Physical Review B* 75, no. 6 (2007): 064205. Used with permission.

Nobel citation said that quasicrystals, while brittle, could rein-force steel "like armor." However, when first commercialized, tiny cracks between crystals called grain boundaries made them susceptible to breakdown. After more than a decade of being shelved for any commercial use, quasicrystals may make a come-back from research coming from the lab of the Iranian American professor Ashwin Shahani at the University of Michigan. In a 2021 paper, Shahani and colleagues show that, under certain con-ditions, small quasicrystals can collide and meld together, form-ing a single large crystal with none of the grain-boundary imper-fections found in groups of smaller crystals. This could lead to an array of new commercialization opportunities.[10]

While quasicrystal applications may seem far off, novel alloys of aluminum are far more likely to gain ascendance in the short term. The rare metal scandium strengthens aluminum in three different ways: grain refining, precipitation hardening, and inhibiting recrystallization, or grain growth. With such defining properties, scandium alloys reduce hot cracking in welds, increase strength in the welds, and deliver better fatigue behavior. Scan-dium increases the recrystallization temperature of aluminum alloys to above 600°C, well above the temperature range of heat-treatable aluminum alloys.

Owing to their high resistance to corrosion, high thermal capacity, and durability, Al-Sc alloys are a preferred choice for use in marine vessels, pipes used in the oil and gas sector, and high-voltage transmission wires. There is also likely to be a major growth market for these alloys in the manufacture of liquid hydrogen tanks, which operate in subzero temperatures. With the "green hydrogen" transition, such tanks may well become more commonplace. Airbus Group APWorks GmbH, Germany, in cooperation with Airbus Group R&D, has developed Scalm-alloy, a high-performance scandium-aluminum-magnesium powder metal alloy designed for additive manufacturing of high-strength aerospace structures. This alloy has exceptionally high fatigue properties, with its specific strength approaching that of titanium; it is twice as strong as aluminum silicon powder, currently widely used as an additive in current materials. Airbus believes that aluminum-scandium alloys could reduce the weight of a large aircraft by 10 to 15 percent.[11] The key challenge with the uptake of scandium alloys is the economic viability of extracting the metal, which is relatively rare and usually extracted as a "companion metal" alongside other primary metals, such as zinc. Economic geologists are constantly considering what quantity of an element is needed within a material to make it viable for extraction, given the costs of production. Such calculations will be crucial for the successful uptake of such exotic alloys.

The Russian company Rusal has developed technology to produce scandium oxide from the much-feared "red mud" that ensues from alumina processing (we will return to the topic of red mud in chapter 8). Efforts are underway to produce an aluminum alloy with a scandium content of less than 0.1 percent, which could be harnessed from the red mud itself, thus creating a synergy between the two metals from mines to markets. Alumina production generates usually between 2 and 2.5 tons of red mud per ton of alumina produced. The content of scandium in red mud may be up to 120 grams per ton, and through a process of carbonization leaching, scandium can be retrieved from the Bogoslovsky aluminum plant.[12]

The role of aluminum in the "green transition" is now well established. According to a 2020 study by the World Bank, aluminum is the single most widely used material in solar photovoltaic (PV) applications.[13] Currently, the metal accounts for more than 85 percent of most solar PV components, from frames to panels. The versatility of aluminum extrusions makes them highly suitable for solar panel frames. Furthermore, aluminum studs can enhance solar panel efficiency, thanks to the material's unique reflectivity properties.[14]

An urgent priority for the green transition is to find means of energy storage that can provide consistent renewable power delivery. Battery metals have thus been on the front burner, for both engineers and policy makers. The ubiquity of aluminum makes battery manufacturing with the metal very attractive, and there has been a rush to develop an aluminum-ion battery. The most promising technology in this regard is emerging in Australia, utilizing the synthetic form of carbon—graphene—as a cathode. The Graphene Manufacturing Group, in partnership with the Australian Institute for Bioengineering and Nanotechnology at the University of Queensland, is working to develop a potentially revolutionary battery that can charge sixty times faster than lithium-ion alternatives. The prototype being developed could also hold three times the charge of the best aluminum-ion batteries currently on the market.

The aluminum-ion cells have the advantage of not using cobalt, nickel, or lithium, which all have complex and often high-risk supply chains. Lithium prices have risen from $1,460 per metric ton in 2005 to $13,000 in May 2021; aluminum prices have only changed from $1,730.00 to $2,078.00 over the same period. Additionally, graphene aluminum-ion cells do not require copper, which costs around $8,470.00 per ton.[15] The key limitation that the technology will have to overcome in order to become competitive is the shelf life of the batteries, which is still lower than the lithium-ion battery. The pilot production and testing plant for the graphene aluminum-ion (G+Al) batteries has commenced operations in Brisbane, Australia, in December 2021. If all goes well, the 2032 Olympic Games, which will be hosted by Queensland City, may be utilizing this technology developed on home turf.

Separate efforts are underway to develop batteries that use aluminum metal and the oxygen in the atmosphere as their electrodes. Aluminum-air batteries have the potential to offer among the highest energy densities of all batteries because the weight of electrodes is lighter than other batteries, and energy densities are defined as the amount of total energy output divided by the weight or the battery volume in units of watt-hour/kilogram or watt-hour/Liter. Energy densities are often used to evaluate the performance of a battery. A related performance indicator is power density, which is the amount of energy density available per second (in units of watt/kilogram and watt/liter). These batteries are part of a larger category of batteries, metal-air electrochemical batteries, which use the oxidation of aluminum at the anode and the reduction of oxygen at the cathode to form what is called a "galvanic cell." The aluminum-air battery is a primary "cell" because its ingredients are consumed; the battery, therefore, cannot be recharged. The aluminum metal, Al, is completely reacted to produce aluminum hydroxide, and the oxygen is reduced to produce hydroxide ions. Copper is used to collect the electrical current and is not consumed in the reaction. Crushed charcoal catalyzes and increases the surface area where the reactions can occur, and a saline solution acts as an electrolyte, which carries the charged ions from one electrode to the other.

Research conducted at the University of Rhode Island revealed that the aluminum-air battery system can generate as much energy and power for driving ranges and acceleration as gasoline-powered cars, and the cost of aluminum as an anode can be as low as US$1.1 per kilogram as long as the reaction product is recycled. The total fuel efficiency during the cycle process in aluminum-air electric vehicles (EVs) can be 15 percent (present stage) or 20 percent (projected), which is comparable to that of internal combustion engine vehicles (ICEs) (13 percent).[16] Furthermore, battery leasing programs can create incentives for a circular economy while making the uptake of new technologies cheaper for consumers.

The COVID-19 pandemic made "work from home" a norm, and while we reduced emissions from commuting, there were latent material flows at stake that we often neglected. In the early days of the pandemic, I collaborated with the filmmaker Alex Tyson to make a five-minute educational documentary on how our material infrastructure helped us adapt to the unprecedented crisis. As schools, universities, Wall Street, and Main Street went online, physical borders started to seem less consequential. In order to work, play, and share together virtually, we need a vast infrastructure that isn't just invisible bits traveling over Wi-Fi—but raw, hard material.

The "virtual" internet requires an underlying physical hardware that is highly material intensive. There are over 380 underwater cables carrying internet communications spanning the globe. These cables cover more than 1.2 million kilometers—that's almost three times the distance from Earth to the moon. Satellites carry a small but crucial share of communication and have their own material needs. The solar panels on these satellites require exotic metals—indium, cadmium, and tellurium—to keep them running. Back on Earth, there is a massive network of internet servers, largely hidden from the public eye. These run 24/7—storing and transmitting data for every video chat, social media post, and email.

And just like your smartphone, every internet server requires an impressive list of materials and raw minerals. Add to this the new demands of blockchain, artificial intelligence, and other computing needs, and we keep adding material and energy requirements to the mix. Perhaps the next phase of our innovation economy around minerals will be to create artificial minerals and related materials that can keep up with the pace of our material demands. Up till now, most of such synthetic materials that have been upscaled for commercial usage have relied in some form or another on petroleum. Plastics, synthetic graphite, and carbon composites have all endeavored to replace metals in a variety of applications, but their reliance on fossil fuels has raised some eyebrows regarding their long-term viability. There are some promising developments with aluminum compounds— ones with a rather peculiar and ominous name—that hark back to a discovery of a few centuries ago.

BOILING STONES

In the mid-eighteenth century, the Swedish chemist Axel Fredrik Cronstedt began to experiment with volcanic minerals. He noticed that rapidly heating a particular mineral produced large amounts of steam, which had presumably been adsorbed onto the material. Based on this, he called the material zeolite, from the Greek word *zéō*, meaning "to boil," and *líthos*, meaning "stone."

Natural zeolites are cooked in volcanic cauldrons, but curious mineralogists have since managed to develop synthetic forms. Their microporous properties act as molecular sieves. Cronstedt had likely been experimenting with the mineral that we now call stiblite, collected from the volcanic islands of the North Atlantic or Iceland. The discovery may have also led him to develop the "blowpipe" method of analyzing minerals. Using a goldsmith's tool and a flame, he was able to use changes in the color of the flame to identify different metals in the salts that comprised the minerals. Nickel was discovered as an element by Cronstedt thanks to its characteristic green flame, and the discovery of at least eleven other elements is credited to this method.[17]

Two years after naming zeolites, Cronstedt published a book in 1758 titled *Försök til mineralogie, eller mineral-rikets uppställning*

(An attempt at mineralogy or arrangement of the mineral king-dom). In this book, which was published anonymously, Cronst-edt proposed that minerals be classified based on a chemical analysis of their composition. He was surprised that others sup-ported his ideas and put them into practice, which laid the foun-dation of modern mineralogy. Using this nomenclature, zeolites mainly consist of silicon, aluminum, oxygen, and have the gen-eral formula $M_xAl_xSi_{1-x}O_2 \cdot yH_2O$, where M is either a metal ion or H^+. The value of x is between 0 and 1, and y is the number of water molecules in the formula unit.

Over the past two centuries, there has been such a prolifera-tion of synthetic zeolites being developed that now every new zeolite structure that is obtained is examined by the International Zeolite Association Structure Commission and given a three-letter designation. Algorithms developed by chemists and com-puter scientists have predicted that millions of hypothetical zeolite structures are possible. However, only 232 of these struc-tures have been discovered and synthesized so far, and they have highly specific properties, which can have commercial applica-tions. The inability to synthesize more of the theoretically plau-sible zeolites may be explained by the focus thus far on hydrothermal methods; other methods involving microwave radiation could also be explored further. In general, zeolites are synthesized by heating aqueous solutions of alumina and sil-ica with sodium hydroxide. Simulations have shown that the discovered zeolite frameworks possess a geometric behavior known as the "flexibility window," whereby the material can be compressed but retain its framework structure—these are the most commercially significant zeolites and, conveniently, also the ones easiest to synthesize.

Synthetic zeolites hold some key advantages over their natu-ral counterparts. Controlled laboratory synthesis allows for more uniform, phase-pure manufacturing. Techniques in the lab also allow for the production of zeolite structures that do not appear in nature. Since the principal raw materials used to manufacture zeolites are silica and alumina, which are by far the most abundant geological resources on Earth's crust, the potential to supply zeolites is essentially unlimited. The most widespread use of zeolites is in water-management systems and for animal-feed purification. The benefits of using zeolites in animal feed that have been demonstrated in lab studies include increased mineral utilization, reduction of heavy metal–induced anemia, and reduction of aflatoxin toxicity (a potent mutagenic toxin released by fungi of the genus *Aspergillus* and specially found in raw peanuts).[18] There is strong empirical evidence on the ability of zeolites to remove heavy metals from water, and hence there has been a presumption that they can do the same for biological systems.

The proliferation of integrative health regimens for detoxification using zeolites without clear clinical evidence is an intriguing example of how human beings can take a property in one context and assume its functionality in a totally different one. Even though there is very limited clinical evidence[19] that zeolites can exhibit the same beneficial properties in human biology as they do in farm animals or water-treatment plants, they are so widely used that medical websites have separate entries for them. Their use by cancer patients and those trying to prevent cancer has become so widespread that the Memorial Sloan Kettering Cancer Institute in New York has a detailed webpage to educate patients about the limitations of their use. What is particularly intriguing is that the centrality of aluminum in these orally consumed health supplements is often not widely noted, even though there is serious concern around the use of aluminum in cosmetics and antiperspirants (as discussed in chapter 2).

Regardless of their health impacts in terms of direct consumption, the most promising future use of zeolites is likely to be in nanotechnology. The porosity of zeolites gives them highly versatile properties at the molecular level of bioengineering intervention. As one recent study noted: "Porous materials are promisingly sustainable solutions for some global concerns including rising energy demands along with the need for stricter environmental approvals for industrial contaminants, depleting resources, and improving health."[20] The market for these unique aluminosilicates is expected to grow considerably in the coming years. However, elementally, aluminum remains hidden in their midst. Massive energy is not needed to extract the metal itself; instead, we rely on its properties of molecular flexibility. Ultimately, if we can find abundant minerals with functional uses involving minimal energy input, this would be the holy grail of material innovation. However, such niche applications as those provided by zeolites are unlikely to be expandable for most uses of aluminum metal.

Metals that form minerals flow through Earth's crust and churn themselves up through volcanic processes, just as zeolites naturally froth through hydrologically exposed lava tubes. There are geological cycles that we can replicate in labs to create zeolites. However, extricating the metal from the mineral gives us an opportunity to accelerate the cycling of a useful material. The linear conversion of a mineral to a metal over geological timescales gives way to the potential for circularity at the level of the metal. Reconciling both these processes will lead us to the ultimate means of attaining some sustainable elemental flows, as we venture into the final two chapters of this book.

IV

ELEMENTAL FLOWS

All those engaged in the extraction, processing, and use of materials must be constantly straining to look into the future as best they can, using the wisdom and intellectual tools that can be mustered from whatever disciplines turn out to have something to offer. This is the nature of the challenge of materials and society. . . . Wider governmental, and eventually multinational intervention in the allocation of scarce materials resources may become necessary to supplement the signals from the marketplace which may not act sufficiently far in advance. On the other hand, there is a growing realization in the underdeveloped world that they control the resources which the developed world needs to feed its industrial machine and can increasingly set their own prices. This may not be wholly bad in that it will tend to stimulate more innovation in the economical use of scarce materials and in the development of substitute materials. Furthermore, in the long run it may provide a more acceptable way of transferring needed foreign exchange to the less-developed countries.

—Harvey Brooks, *Materials in a Steady State World,* 1972

7

RECYCLING AND REALISM
The Industrial Ecology Paradigm

For a successful technology, reality must take precedence over public relations, for nature cannot be fooled.
—Richard Feynman

Growing up in Lahore, Pakistan, in the 1980s, my first encounters with the world of recycling were visits every Friday by the local scrap-metal dealer, who would ride around the neighborhood on his bicycle. Two huge sacks attached to a pole were precariously balanced on a carrier above the bicycle's rear wheel. He would make a loud call, asking for metal scrap as he wobbled through the streets to keep his bike's balance while maintaining a slow enough speed to keep an eye out for potential customers. Old pots and pans, metal wires, and a wide assortment of tin cans and knickknacks would be brought to him by residents. He would take some for free, and he might make a nominal payment for others. I recall a notable absence from his repertoire, from what one might expect in a recycling pickup today. No aluminum cans were used for beverages in Pakistan at that time. The multinational soda companies like Coca-Cola and PepsiCo had a presence in the country but filled their products in reusable glass bottles, which were washed and sanitized rather than crushed and recycled.

Glass bottles have endured the threat from many metallic innovations. Tin cans for food storage were patented in 1810 by the English inventor Peter Durand, who used a tin-coated iron can, which could preserve food after heated sterilization. Tin was a rare metal in comparison with iron, but its main attraction for this purpose was its resistance to corrosion by food acids. Furthermore, tin forms bonds with other metals in ways that made coatings highly effective. Tin is also biologically inert in humans and nontoxic in most forms, making it particularly suitable for can coatings. The Royal Navy was an ideal market for this invention because sailors needed provisions over long uncertain distances on the high seas. Well-preserved canned food enabled more ambitious expeditions. Explorers attempting to navigate the Northwest Passage across the Arctic Ocean, such as Otto von Kotzebue of Russia, wrote of the importance of the metal can's role in provisions for his 1815 voyage. The Welsh explorer Sir William Edward Parry made two Arctic expeditions in search of the Northwest Passage in the 1820s and took canned provisions on his journeys. A four-pound tin of roasted veal, carried on both trips but never opened, was kept as an artifact of the expedition in the Royal United Services Museum in London until it was opened in 1938. The contents of this century-old entree were chemically analyzed and found to have kept most of its nutrients and to be in good condition. Although some heat-resistant bacteria had infiltrated the can, the veal was fed to a cat, who had no complaints whatsoever.[1]

Despite their popularity among the explorers and the military, metal cans would not enter commercial use until the mid-nineteenth century, when Henry Evans granted the patent for the pendulum press. This invention, in combination with a mechanical die, accelerated production from five or six cans per hour to fifty to sixty per hour, inaugurating the age of commercial canning. However, most of the cans made from then onward until the middle of the twentieth century were discarded after their use. While scrap metal has been recycled since antiquity as a natural human response to the scarcity of ore, large-scale commercial production disincentivized the recycling of many metallic goods. The abundance of iron and its most popular alloy, steel, made mining, smelting, and refining metal from ore more attractive than recycling the earlier generation of metal cans. All this changed with advent of the aluminum can—which would eventually become the most recycled of all products.

William K. Coors, of the eponymous beer company in Golden, Colorado, is credited with "inventing" the aluminum beverage can after four years of research in 1959. In its announcement of this invention in the *Colorado Transcript* that year, the company explicitly stated that recycling was a motivator for this invention—noting in the secondary heading that "metal in can is reusable."[2] Commercialization of the aluminum beverage can was started by the Reynolds Metals Company in 1963 and used to package a diet cola called Slenderella. Royal Crown adopted the aluminum can in 1964, and by 1967 Pepsi and Coke followed. A key innovation for the packaging industry from the usage of aluminum also came because the aluminum can could be manufactured from only two pieces—a body and an end. This was possible thanks to the malleability of the metal. The design also made storage of carbonated sodas and their transport much more secure, with less likelihood of spillage. By 1985, the aluminum can was the most popular beverage package in any market. But the uptake of aluminum in the beverage industry of developing countries took much longer.

The absence of aluminum cans in the repertoire of the bicycled scrap dealer that frequented our Lahore neighborhood in

1980s reflected a set of tradeoffs that are inherent in the uptake of what we now call a "circular economy." Bottles could be washed (if there was abundant water available) and reused, but they were heavier and fragile. Recycling of scrap requires infrastructure for its collection and reprocessing into usable material. And then there is the economic tradeoff between all these options and just getting raw materials and developing a virgin supply chain. Making all these calculations necessitated the emergence of new metrics and techniques, which ultimately became incorporated in the 1990s into the academic field of industrial ecology. As the name implies, this field recognizes that industry is now a permanent feature of the ecological systems on the planet and that the inputs and outputs of industry need to be incorporated into a paradigm that resembles natural cycles. How such cycles can be emulated in ways that are most economically and energetically efficient underpins a fundamental quest for sustainability in the Anthropocene. Such a quest for cyclicality will need both public policy and an entrepreneurial spirit to find new technologies but also a new generation of industrialists who see the virtue and value of circular material flows in the economy.

THE MATERIAL DIASPORA

Gazing at the vast expanse of buildings from any skyscraper window in Dubai is humbling on multiple levels. There is the sheer scale of construction material employed to build this metropolis between the desert and the sea. Look closer, and there is an even more remarkable revelation. This city, along with numerous others in the fossil-fueled wealth of the Arabian Peninsula and its geological environs, hosts the largest diaspora community of migrants per capita. They have brought their talents from far and wide to provide not only labor for the construction of these massive cities but also innovation and cosmopolitanism. Notwithstanding the valid criticism of the difference between labor and citizenship rights, the massive material economy has brought highly talented entrepreneurs to the region as well.

Vinay Kumar Dewan exemplifies such an entrepreneurial diaspora spirit, one that transcends the linear, extractive mindset of the Gulf region and incorporates industrial-ecology approaches in the most unlikely of industrial establishments—an aluminum smelter. The personal story of his family exemplifies how minerals extraction has a visceral impact on entire lineages of immigrant families. Even when they leave their lands of origin, immigrants often continue to maintain a bond to material professions. The Dewan family's story is like many material diasporas that one finds across the world. For example, the Eastern European Jewish community who migrated to South Africa, Belgium, and Israel carried with them their prowess in diamond supply chains and the craftsmanship needed to cut that hardest of gems. The Dewans traced part of their ancestry to my home city of Lahore as well but had settled in Batala, in what is now northwestern India, after the Great Partition of the subcontinent by the British colonialists. Vinay's father H. R. Dewan was among the first class of graduates from the Indian School of Mines, which had been established by the British in 1926 in partnership with the Royal School of Mines in London. At the time, the most important mining commodity was coal, and the location for the campus was chosen to be Dhanbad—a city in the present-day Indian state of Jharkhand, which is considered the coal capital of the country.

H. R. Dewan had always been more interested in metals and saw the potential for a much broader industry around them than mining coal, which was eventually only destined to be burned.

His career took on the base-metals sector in India, which was largely controlled by state enterprises at the time. Two of his four sons also embraced mining as a career and graduated from the Indian School of Mines. Vinay Dewan was the younger of these two sons. He was born at a copper mine near Alwar, Rajasthan, during his father's posting there. He became deeply interested in the bauxite boom of the 1960s in India and recognized the coal-aluminum nexus: the massive energy requirements of bauxite processing and refining led to a natural linkage between the two sectors.

When I interviewed Vinay Dewan at his apartment in a Dubai skyscraper in November 2021, his memory of the finer details of materiality was impressive. The family car during his childhood years was a Willys station wagon—which had an unusual wood and metal frame. His fascination with the components of our material lives was also inspired by a physics teacher in his youth named Mr. Saifuddin. Science spurred an inquisitive impulse that transcended the ethnoreligious differences between the Muslim teacher and his erudite Hindu pupil. During those years of the Cold War, he was also influenced by cinematic representations of nuclear weapons and their material impact. Across from the Bishop Cotton Boys' School in Nagpur, where Vinay studied, was a movie theater, where he first saw a Japanese film called *The Last War*, which depicted a dystopian future led by an arms race that could be linked to a military-industrial complex. The material profligacy of the defense establishment was portrayed in this film, as was the destructive power of human ingenuity.

As a student at the Indian School of Mines, Vinay Dewan wanted to take courses in explosives engineering, but none were available. He managed to get at least eight students interested in the subject to recruit the chief scientist of the Central Mining

Research Institute of India to teach the course. Specializing in explosives gave Vinay considerable marketability with the defense establishment, but he chose to apply his learning to mineral extraction. During the 1970s, when the Gulf States began to encourage foreign investment, he spent time in Iraq, Iran, Kuwait, and the Emirates. During his travels, he even met Saddam Hussein before his full ascent to power. Initially, he applied his explosives training to the quarrying of limestone in the emirate of Ras-al-Khaimah. Calcium carbonate from limestone had many chemical applications but was also indirectly useful in the processing of metals.

Material entrepreneurs are most successful when they can make connections between industry supply chains and find points of leveraging efficiency. From his training in metallurgy, Vinay Dewan was aware of how extracting aluminum from bauxite ores required alkaline interventions. In addition to the Bayer process, which utilized sodium hydroxide to leach out aluminum oxide (alumina), there is also the use of soda ash (sodium carbonate) in certain resistant ores. Limestone is the primary geological resource to produce soda ash, and hence a connection between a quarry of calcium carbonate derived from ancient organic shells can be made to the production of aluminum. Vinay Dewan saw this connection but also recognized the major energy minerals of the Gulf States, which could be used to subsidize power for metal smelters. Thus, when an opportunity arose, he directed his business acumen toward linking the quarrying and blasting businesses to broader metal supply chains.

Around the time that Vinay Dewan was contemplating his Gulf enterprises, the tiny island state of Bahrain won independence from the United Kingdom in 1971. The king, or amir, of the island state chose not to join the United Arab Emirates or its neighboring peninsula country of Qatar, with which he had historic tribal rivalries. Instead, Amir Sheikh Isa bin Salman Al Khalifa chose his own hybrid brand of governance, in which foreign investment and modernization was encouraged long before Dubai came to the world stage. He was concerned with diversification of the country's economy and chose to invest in an aluminum smelter that could utilize the cheap energy from access to oil and gas to establish another export commodity. Hence was born the Aluminum Company of Bahrain (ALBA), which started operating the first aluminum smelter in the Gulf, producing 120,000 tons of metal concentrate. Meanwhile, Vinay Dewan had seen opportunity in Bahrain's open financial system and started to operate the only quarry on the island, which began operations in 1976. Over the next decade, Dewan won the trust of the Bahraini government and quietly observed the growth of the metals sector.

By 1990, after twenty years of production ALBA had accumulated over 30,000 tons of waste material at the site where new infrastructure was to be constructed.[3] By then Dewan was leading the BRAMCO Group—the only mining company on the island—and saw an opportunity of removing and recycling this waste stockpile. He had a strong commitment to ecological management, and the complexity of aluminum's production process did not daunt him but rather provided an opportunity for a range of product streams from the ostensible "wastes," which included:

- Broken cathodes (graphitized carbon)
- Broken anodes (calcined pet coke)
- Tar pitch (used in paste plant of alba)
- Collector bars
- Mixed quantity of solid bath (tapped cryolite)
- Cast iron from anode butts

- Empty drums
- Timber, metal, tools, hand gloves, etc.
- Spent pot linings consisting of silicon carbide
- Bricks, refractories, etc.

BRAMCO drilled about one hundred holes diagonally under the stockpiles around its periphery. Samples from holes were drawn at every two meters of depth, up to eight to ten meters. Although toxic leachates were not found under the surface, there were undoubtedly going to be some possibly hazardous residual materials left after the salvaging of useful materials. The company developed a cold process to recover aluminum metal from the waste, or "dross," via electromechanical processing. The dross was crushed to 10–12 mm size and passed over a powerful magnet that removed any iron particles. By means of an Air Cyclone suction system, fine particles of alumina were removed and bagged separately, to be sold to the ceramic and glass industry or palletized and sold for slag conditioning to steel plants. The remaining product was passed over eddy-current linear motors, which bounced out the metallic particles.

The metallic components could then be tumbled in a ball mill, using silicon carbide from the smelter itself as the grinding media. Abrasive silicon carbide could clean the granules of any oxides adhering to it, and shiny aluminum granules of spherical shape were obtained and sold directly to the steel mills as deoxidation granules. This proved to be the most efficient way of processing dross without much heat input or pollution. Dewan's process reduced the carbon footprint, and the yield was maximized with minimal power input and no burning loss. The other solid materials were screened at site and analyzed for their toxic contents. There were minor traces of cyanide and fluoride in the smaller particles, and all fractions below 25mm were considered semihazardous. This residue was then disposed of in a specially insulated dumpyard.

Dewan, motivated by his young daughter Kanika's vision on sustainability, soon became the recycling magnate of Bahrain, and he expanded his operations to various other material sectors. He called his new company Scrapmould, which committed itself to not just recycling but "upcycling" products to more productive uses. Ultimately, this is the gold standard for circular material flows to ensure minimal planetary impact in terms of energy consumption and management of "entropy" (chaotic distribution) of materials. Circularity of material flows became a personal passion, and in his spare time he started to find ways to construct furniture and art from waste materials. When I visited the family apartment in Dubai, the dining room was adorned with a huge piece of artwork crafted out of waste wood fragments and stools made from old copper scrap and equipment. His daughter Kanika started the Zero Waste Art movement in her own professional work as a building designer and artist. Many of the lessons from Bahrain worked their way back to the Dewans' Indian homeland. In the early years of the current millennium, Kanika Dewan won the contract for all the stone works of the two largest airport projects in India: Terminal 3 at Delhi's Indira Gandhi International Airport and the new Chhatrapati Shivaji International Airport in Mumbai. She utilized recycled materials as much as possible in the flooring despite the insistence of public officials that the term "waste" not be used in describing any of the inputs. Her quest for zero waste also transcended material "silos" when she innovatively designed and produced the first-ever lightweight junction boxes in Delhi International Airport from waste stone, backed by waste aluminum panels. The material diaspora is well attuned to the globalized flow of complex supply chains, yet cultural stigma toward "waste" must still be endured. Eventually, entrepreneurial talent can blossom across what the United Nations International Resource Panel terms "material cascades" (as shown in Figure 7.1).

The Dewan family's tale of material innovations across a range of industries is emblematic of the way smart industrial entrepreneurship can create value in unlikely locales. The positive impact of this "material diaspora" is spurred by the productive channelization of capital though a cavalcade of industries and the power of globalized learning. Yet the challenges of collecting materials from the average consumer back into the recycling system remain. A separate set of tactics are being employed to manage such sustainability challenges.

THE HYDRO-METAL NEXUS

The flow of metals through the economy is analogous at some level to the flow of water in the hydrological cycle. There are "reserves of water" in the oceans, icecaps, and aquifers—some of which are more accessible than others. Water is even locked in crystals within rocks—in some ways similar to how elemental metals are locked away in minerals. A clay mineral known as smectite could hold a substantial portion of the water "missing" from Mars. If we consider water trapped in such minerals, the Red Planet could have held enough water to cover its entire surface in a layer one hundred meters and one kilometer deep![4] However, the connection between water and minerals is one that harks back to the second law of thermodynamics. Metals trapped in minerals or synthetic products can be in a higher state of entropy and require energy to extract them in pure form. Flowing water can provide that energy in ways that would make that extraction more efficient from both an ecological and economic

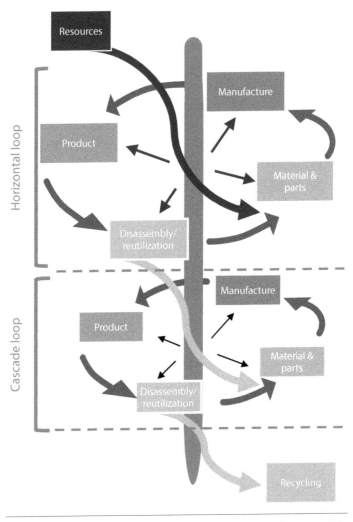

Horizontal loop

Cascade loop

Resources

Manufacture

Product

Material & parts

Disassembly/ reutilization

Manufacture

Product

Material & parts

Disassembly/ reutilization

Recycling

FIGURE 7.1 Potentials for entrepreneurship within a cascading material value-retention system.

Source: United Nations International Resource Panel, open source.

perspective. Hydropower and aluminum extraction have had a close connection since the early days of Alcoa's properties near Niagara Falls or the Saint Lawrence Seaway. The same was also true of European smelters that gained traction in the sparsely populated regions of Scandinavia. The origin story of what is now one of Europe's largest aluminum companies and one of the world's most efficient recyclers of the metal shows us how reliable hydroelectric energy sources can lead to industrial inventiveness and the "reinvention" of product streams.

Sweden was known for its industrial prowess at the turn of the twentieth century. Despite the country's current claims of socialist egalitarianism, there has been immense control of wealth within the hands of a few industrial elites in Sweden. In particular, the Wallenberg family held immense power with the country's industrial investments and was also responsible for developing metal-processing capacity in Northern Europe. At a time when Norway was still in a tenuous union with Sweden, the Wallenbergs established Norsk Hydro-Elektrisk to use hydropower to make nitrogen-based fertilizer. By the 1920s, when Norway had become an independent country, Norsk Hydro's electric-arc-based technology for manufacturing artificial fertilizer was superseded by the German Haber-Bosch process for nitrogen fixation and ammonia production.

As the nuclear age dawned, the company transitioned to using its energy supply to manufacture "heavy water" (in which the hydrogen atom had an extra neutron, forming the isotope deuterium). Deuterium was used in nuclear reactors but also gained uses in chemical-analysis techniques such as nuclear magnetic resonance (NMR). Interestingly, deuterium under extreme conditions can literally merge the "hydro-metal" nexus, exemplifying how illusory the physical properties of matter we assume inhere in each state are. A substance like hydrogen, which we associate as a light gaseous nonmetal on Earth, can exhibit metallic properties as well. In the interiors of giant planets such as Jupiter and Saturn, hydrogen's heavy isotope gets squeezed until it becomes a liquid metal capable of conducting electricity. Under about 1.5 million times normal atmospheric pressure (150 gigapascals), the deuterium switches from transparent to opaque, absorbing the light hitting it instead of allowing it to pass through, and at nearly 2 million times normal atmospheric pressure (200 gigapascals), metal-like reflectivity starts.[5]

Once the deuterium market became more limited and a greater scrutiny of the sales from Norsk Hydro to Germany during the Second World War began, the company started to move its energy interests toward light-metal production. In 1940, construction of a magnesium carbonate plant started at Herøya, but the German invasion of Norway halted the plans. Since 1919 there had first been zinc and then aluminum production at Glomford—a small industrial town just north of the Arctic Circle. With a range of other industrial endeavors hitting dead ends, Norsk Hydro eventually realized the synergy between hydropower and aluminum production was a winning proposition. Within a few decades, the company broadened its aluminum exposure, with aluminum-production acquisitions from Brazil to Qatar.

The high energy footprint of aluminum processing from mined ore has naturally created an economic incentive for greater recycling. However, Norsk Hydro saw the potential to leverage its hydropower advantage to aim for carbon neutrality, as well. With concerns growing globally about climate change, corporate and government attention on metrics of carbon emissions worked to Norsk Hydro's great advantage. The aluminum sector is responsible for 2 percent of the greenhouse-gas emissions from human activities, according to the International Aluminum

Institute. Most of the sector's emissions are produced during smelting, which requires an enormous amount of electricity that, in many cases, is generated from fossil fuels. Norsk Hydro is developing carbon-free aluminum-smelting technology that converts alumina to aluminum chloride before electrolysis, with an industrial-scale pilot planned before 2030. Norsk Hydro has estimated that with its combination of magnetic and X-ray-separation technologies, around 97 percent of all aluminum scrap is in fact recyclable.

The company has started to market Hydro CIRCAL as a "premium recycled aluminum" with a minimum of 75 percent recycled, postconsumer aluminum scrap. They also offer a "Hydro RESTORE" brand of aluminum products made from a combination of recycled preconsumer scrap, recycled postconsumer scrap, and primary aluminum. The main ingredient in Hydro RESTORE is preconsumer scrap, which is generated in the production process at existing aluminum production facilities. Norsk Hydro collects this material and recycles it, so it can return as aluminum for new products.

The link between water and aluminum recycling goes beyond just the metaphor of flow. Optimizing a circular-economy approach requires project developers to keep track of material flows and compare impacts of a range of output options. The field of industrial ecology provides a range of tools that should be utilized in this regard. First, there must be clear accounting methods for energy and material flow, which can be tracked in direct terms as well as in impact metrics such as carbon footprints.[6] Once we have material and energy flows delineated, lifecycle analysis (LCA) techniques can be used to track the environmental and social impact across the full supply chain, up to the use and disposal of a product. Such an approach provides a comparative impact matrix that considers the relative systems-wide impact across the full supply chain as well as the disposal and recyclability impacts.[7] Among the most challenging aspects of such an analysis is considering the optimal "life" of a product. While durability is overall a positive attribute for products, the potential for innovations that could lead to efficiencies in future products needs to be considered. The question of when to "dispose of" a product, to optimize its ecological and economic efficiency, is thus one of the most difficult dilemmas for industrial ecologists to consider.

THE DURABILITY DILEMMA

Located on both sides of the Benin-Niger interstate highway, eighty-seven miles from Cotonou (the largest city of the small West African nation of Benin), is the village of Cana—now popularly known as the "Aluminum Village." Cana was once considered a holy city when King Agadja built his palace here in 1708. From his reign onward, every new monarch enthroned at the main palace in Agbomê was required to erect a secondary palace at Cana. The village also derived its prestige from the establishment of the dynasty's great protective voodoos and was once a celebrated spot in the Danxomê kingdom, which derived its wealth from the slave trade. Cana lost its wealth and charm after the annexation of the area known today as the Republic of Benin into French West Africa, the culmination of a process the Senegalese writer and philosopher Cheikh Hamidou Kane described in his prizewinning novel *L'aventure ambiguë* as "the art to vanquish without being right."[8] So how did this once imperial town get its current metallic title?

In the postcolonial era, one of the clans of Cana began to see value in collecting metal scrap from a nearby dump and using a rudimentary earthen furnace to melt the aluminum into utensils and divine figurines. In particular, the villagers worship Asin—a god they believe was crafted out of aluminum by their ancestors because of the metal's ability to deliver livelihoods to the village. The villagers now consider aluminum carving a part of their cultural identity, even though it is at most two generations old. There are voodoo sacrifices of cackling chickens and prayer "gargles" from priestesses made before the carved aluminum figurines of Asin, which sell quite well on a regular basis alongside the utensils. When the filmmaker Onesiphore Adonai documented the remarkable story of this village for an Al-Jazeera documentary in 2021, he also discovered that the youth were very concerned that there was now less aluminum to be found in the local dump. A steady stream of scrap was essential for the livelihood of this village, but for that to occur, disposable products needed to make it to the dump.

Durability is thus inextricably tied to livelihoods, not just for the manufacturer but also for the waste picker. This is not just an issue for the developing world, because much of the manufacturing happens in developed countries. The dilemma confronting industry even received Hollywood attention in the 1950s satirical comedy film *The Man in the White Suit*, starring Sir Alec Guinness as Sidney Stratton, a mercurial chemist who synthesizes a textile that's resistant to dirt and degradation. While a beacon for more sustainable attire, this fabric spreads anxiety in a Lancashire mill town whose population depends on a flow of wearable—and tearable—garments. Unions and corporate bosses unite against the humane idealism of Stratton. The movie represents the filmmaker Alexander Mackendrick's genre of cinema, which has been termed "lethal innocence." In addition to capturing the perils of naïveté in a hyperconsumer society for the innocent chemist, the film also captures the challenges of a phenomenon that economists call "planned obsolescence." As the term suggests, there can be an economic incentive to ensure that a product becomes obsolete after some time to allow for new products to enter the market. This can be motivated by a desire to have the most innovative products; it can also be motivated by a crass desire for throughput.

Increasingly, it is becoming apparent that many large corporations are deliberately designing products for planned obsolescence, which is having a massive impact on material flows—especially of metals. In 2019 alone, by one estimate 50 metric tons of electronic waste were generated globally, with only around 20 percent of it officially recycled. Half of the waste was from large household appliances and heating and cooling equipment. The remainder was TVs, computers, smartphones, and tablets. The problem has attracted attention by civil society groups as well. In 2020, Apple agreed to pay up to $500 million to settle litigation accusing it of deliberately reducing performance of its gadgets and thereby purposely rendering them obsolete. The class-action lawsuit makes the company pay $25 per iPhone in damages. While the company denied wrongdoing, it claims to have settled the nationwide case to "avoid the burdens and costs of litigation," court papers show.[9]

The ability to maintain product life and retain associated jobs and livelihoods requires us to shift from a linear manufacturing model to one coupled with service-sector employment in repairs, remanufacturing, and recycling. Such a shift requires us to appreciate the intricacies of what is called a "complex adaptive system." Note that "complexity" in such a context is different from being "complicated." An automobile assembly plant is complicated but not complex. *Complicated* systems may have numerous parts and connections, but their outcome of operations is largely predictable. However, *complex* systems, such as a circular economy, may have a low or high number of paths and connections, but their nonlinearity, feedback loops, and emergent behavior make them highly unpredictable. Figure 7.2 shows an example of how we might consider the durability dilemma in the context of materials such as metals or plastics.

Starting from the left, the diagram lays out key intervention points for a circular-economy paradigm ranging from research areas in "green chemistry" and recycling technologies to improving scrap and waste-collection systems. Actual material-needs assessments are also an essential component of any project design in a circular-economy context because "dematerialization" is always a desired goal in terms of aggregate reduction of consumption. Promoting upcycling of materials rather than downcycling where possible as well as waste-to-energy conversion projects ensures maximum resource-flow circularity wherever possible.

The circular-economy paradigm connects with a range of policy experiments and visioning exercises that are now being carried out worldwide. For example, the Australian government developed a Resilience, Adaptation, Pathways, and Transformation Assessment (RAPTA) protocol for "Sustainable Futures."[10] However, such complex systems can be highly sensitive to any perturbations, and monitoring must be a key component of ensuring that positive transitions from linearity to circularity are sustained. For circular-economy projects, material-flow accounting and lifecycle analysis are important tools for the monitoring and evaluation phases. Keeping track of total consumption reduction is also important to avoid what is called the "rebound effect."

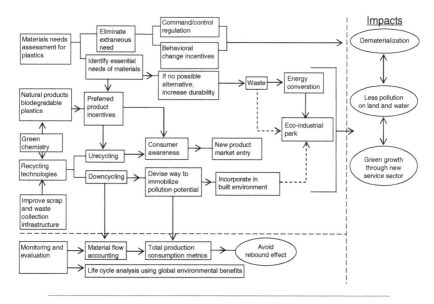

FIGURE 7.2 How change can be galvanized toward a circular material economy.

Innovation to improve efficiency can be an important feature of finding win-win outcomes for a circular economy. However, there is a feature of material and energy economics termed "Jevons' paradox" (after the nineteenth-century coal economist who proposed it) that we must be aware of. The paradox arises from an observation made by Jevons that as energy technologies became more efficient, they did not always lead to reduced consumption, which one would expect. In fact, efficiency might paradoxically lead to *higher* consumption. This paradox manifests because if there is latent demand for consumer goods, efficiency within a circular economic system could still increase overall consumption. In other words, since efficiency is likely to reduce production costs, this can lead to lower prices, which can increase the demand for the good, causing a "rebound" in consumption. Also, if consumers think a material is more sustainable in terms of its ecological footprint or its resource efficiency, they may feel more comfortable consuming more products with it than they might have otherwise, possibly negating the benefits of its lower impact metrics. The observation is often presented by environmentalists as a counter to technological optimism about energy and material efficiency. The same argument can also be made for "dematerialization"—as less material is used for manufacturing one product, more of the raw material will be available for other goods, possibly spurring latent demand.

The concept of a circular economy is a definitive paradigm shift in the way industrial processes relate to the modern economy. If we are monitoring systems-level consumption patterns, the rebound effect stemming from Jevons' paradox can be avoided. The conventional economic model has been focused on linear material flows from mines to markets. However, a circular-economy approach suggests the need to reconfigure economic systems around finding ways of conserving energy and material flows through cyclical processes that mimic natural cycles in ecosystems. When we refer to "systems thinking," it is important to consider four of the key characteristics that define a system in this regard:[11]

1. *A system's components are not a collection of elements but are interconnected to and affect one another.* Example: A transport system in a city is not just an additive assemblage of the number of buses or trains but how those transit nodes interact with one another and with travelers.

2. *All the components in a system get organized (either by design or by emergent evolution) in a specific manner to achieve the system's goal.* Example: Fish in a shoal behave as a system and can self-organize to distract predators, a behavior that is a product of emergent evolution.

3. *A system can have a specific function in a larger system wherein it becomes a "subsystem."* Example: A pond with its assemblage of aquatic plants and animals is a system that is self-contained, but it also interacts with the local hydrological environment and the atmospheric system that comprise the ecosystem.

4. *Systems have feedback loops that can be negative or positive in reinforcing or mitigating trajectories and transitions.* Example: In the climate system, volcanic eruptions can discharge sulfur, which can create a negative feedback loop for global cooling in what may otherwise be a locally warming event.

In many ways, the circular-economy narrative recognizes that nature has developed time-tested systems of biotic innovation that should be mimicked by human systems. Thus, the human industrial system should be intertwined and complementary with the natural system. Circular-economy approaches are therefore synergistic with what conservationists are calling "nature-based solutions" as well as offering greater efficiency.

A neglected aspect of circular-economy research has been an evaluation of how such a paradigm would affect basic human development challenges. There seems to be a presumption that "win-win" outcomes would emerge from efficient systems in a circular economy that could provide development dividends in the world's poorer nations. Yet some of the dominant premises of a circular economy necessitate reduced consumption and the increased durability of material products. This in turn has the potential for a major impact on human development in areas that depend on livelihoods from those processes. As a locus of analysis, the consumption of myriad products and services provides an essential link between economic development and environmental impact. Yet a polarized view that considers consumption as only an ecological *problem* of human decadence needs to be avoided. Optimists in this regard would argue that a transition to a service sector and its concomitant wealth creation would counterbalance the reduced throughput of manufacturing employment and livelihoods for market economies. The transition of livelihoods following the automation of major labor-intensive industries during the past century is often alluded to in this vein. However, there are limits to the absorption of employment by the service sector in densely populated developing countries such as Nigeria, which are aspiring for major development outcomes for their population beyond aggregate indicators of growth.

Within the context of a circular economy, it is also important to note that subsistence economies where people grow their own food or are completely self-reliant have their own challenges toward realizing the goals of sustainability. Although subsistence societies have important survival skills for meeting basic human needs for food and shelter, they are too focused (by necessity) on their immediate family or tribe to contribute meaningfully to broader societal ambitions. The innovations emanating from subsistence societies are focused on a very narrow sphere of influence. The villagers of Cana in Benin exemplify how a subsistence economy has bridged to a market system within a traditional cultural context while relying on circular material flows of a metal like aluminum. Transitioning to a circular economy also requires us to consider consumer choice, which can't just be "engineered," as an important part of the model.

POLICIES BEYOND
PLANNED OBSOLESCENCE

Within a circular-economy model premised on adaptive learning from a "socio-ecological system," we can likely escape the efficiency trap. As the term suggests, such a system connects social science with ecology. While Jevons' paradox may well hold in areas with growing demand, it is less likely to apply in mature markets. For example, in a developing economy energy efficiency and reduced cost may lead people to buy more products to use more energy. However, in a market with a fairly saturated consumption profile (one without many more things to add to material and energy demand), efficiency and dematerialization are very valuable (figure 7.3). We can thus be cautiously optimistic

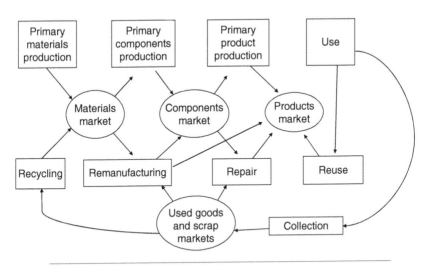

FIGURE 7.3 Simplified policy interventions to prevent planned obsolescence.

Source: Adapted from T. Zink and R. Geyer, "Circular Economy Rebound," *Journal of Industrial Ecology* 21, no. 3 (2007): 593–602.

about the prospect of channeling our treasure impulse toward seeking efficient material usage.

Planned conservation and the efficient cycling of wastes may extend our time horizon for depletion and perhaps give technology a greater opportunity to find alternatives. Ultimately, our only salvation in grappling with the durability and development dilemma will be in finding energy sources that can be developed most effectively with existing materials available within a circular economy. One of the key tenets of the philosophy of "dialectical naturalism," coined by the American environmental philosopher Murray Bookchin, is that incremental change leads to "turning points." The term is a play on the philosophy of dialectical materialism developed by Marx or the dialectical idealism of Hegel. The term "dialectical" implies a conversation to resolve contradictory goals through feedback loops. However, instead of conversations between ideals or between market philosophies, the dialogue implied here is between society and ecology. The long-term viability of a circular economy will ultimately depend on a turning point around the availability of efficient and effective energy. Such a turning point would allow for many of the other contradictions and imponderable elements of reconciling material flows, and development challenges would also be addressed. However, there would need to be a global effort to ensure resource flows are most effectively channeled in this regard through improved governance systems.

There has been a proliferation of civil society organizations in recent years that are particularly concerned about the impact of electronic devices on the environment and society. The development of smartphones and myriad other devices that are connected to the internet has led to a vast genre of research into ways of measuring and mitigating their adverse impact. The British charity Green Alliance brought together a range of research on policy efficacy at the corporate and governmental level to suggest six key pathways for reform. They also considered the viability of these policies in three different policy cultures: the United Kingdom, the United States, and India. Among their key findings was that software is now just as important as hardware in reducing material-flow impacts and planned obsolescence.

Twenty percent of consumers say that they replace a device when there are no more software updates for the old one. The circular-economy strategy to address this challenge can follow two avenues: the first would be simply to have a longer warranty support period for the product. The second is to develop better firmware and provide an additional "keep your device running quickly" service for existing users for free or for a small upgrade fee. Another software intervention pertains to better saving opportunities beyond the device, for example, "cloud offloading." Services could be sold that bundle access, device, and performance together. The main benefit of cloud offloading may be to lower cost. Less-established manufacturers could focus on durable, lower-end new and remanufactured devices rather than going into direct competition with manufacturers who are competing on high-cost, high-end new devices.

Such efforts could be coupled with better reuse options for old devices that lie abandoned in drawers—estimates from the Green Alliance's work suggest that one-third of all cell phones ever made are currently in such states of material limbo. The key to maximizing the value and environmental benefits of reuse is to retrieve used devices as early as possible, although there is value in many devices for over five years from the original sale date. Used devices see immense variation in the cost of minor repair: many devices don't have easily removable batteries or replaceable screens, as designers favor slimness over repairability. Modularity is thus another important feature of policies that would allow for a more sustainable material flow. Remanufacturing and parts harvesting would extend the minor modularity envisaged in model 3 in figure 7.4 to include improved "disassemblability" and compatibility of additional components. This would enable the reuse of components when the device is otherwise not able to be repaired. Such an approach would keep high-embodied-carbon components, such as integrated circuits, which contribute 35 percent of a smartphone's carbon footprint, in use for longer. Finally, opportunities for do-it-yourself repairs are important and gaining traction with sites like IFixit. Redesigning devices for repairability and providing information on how to repair them would enable customers to address these problems themselves. However, many IT manufacturers have resisted such an approach: Toshiba has refused to release its repair manuals, citing intellectual property rights, and Apple has developed proprietary screws to prevent customers from opening their devices. Such barriers by industry require regulatory action for change. Figure 7.4 presents a summary of these interventions and their process for modulating impact.

Despite the pandemic, toward the end of 2020 the European Union took some ambitious steps to address the more than 12 million tons of electronic waste the bloc produces annually. Acknowledging that "Europe is living well beyond planetary boundaries," a European Parliament vote called for mandatory repairability scores for consumer electronics, among a host of other initiatives intended to extend products' life spans. Fortunately, there have also been significant advancements in the upgrading of smelters to accommodate e-wastes. The Swedish mining firm Boliden's Rönnskär copper smelter, located outside Skellefteå, has been recycling various waste materials since the 1960s. Today, the smelter's annual capacity for recycling electrical material is 120,000 tons, including circuit boards from

	United States	United Kingdom	India
1 Software-led longevity	Enables growth in a saturated market	Enables growth in a saturated market	Enables redistribution of U.S. and UK devices in India
2 Better reuse	Accesses a large pool of reusable premium devices	Accesses a large pool of reusable premium devices	Enables cascaded use of U.S. and UK devices in India
3 Minor modularity	Addresses common hardware failures and uses existing logistics	Addresses common hardware failures and uses existing logistics	Addresses common hardware failures; may require new logistics
4 Cloud offloading	Dependent on the availability of fast data networks	High population density and good network infrastructure	Niche use only
5 Parts harvesting and remanufacturing	Requires device exports for remanufacturing	Requires device exports for remanufacturing	Enabled by hardware standardization in budget devices and low labor costs
6 DIY repair	Attractive to some consumers	Attractive to some consumers	Builds on existing infrastructure and practice

FIGURE 7.4 Six circular-economy models to deal with metal wastes from "smart" consumer electronics.

Source: Benton Dustin et al., *A Circular Economy for Smart Devices* (London: Green Alliance, 2015). Open source.

computers and mobile phones that are sourced primarily from Europe. The smelted material, known as black copper, then joins the facility's main smelter flow for further refining to extract copper and precious metals.

The smelter has been designed to mitigate the release of potentially hazardous emissions through wet gas purification and other proprietary processes. To abide by the Minamata Convention on mercury pollution, the smelter is also equipped with an additional purification stage for mercury. A key innovation in the design relates to the material-energy nexus whereby plastic in the electronic material melts during smelting, which acts as a source of energy and generates steam. The steam is partially reused for heating the plant, and the remaining heat is supplied to the local district's heating system.[12]

With such technologies in place, numerous nations have made tangible policy interventions: the United Kingdom agreed to enforce EU repair rules, France launched a "repairability index" for select electronics, and Austria reduced taxes on small repairs. COVID-19's impact on consumers also appears to have made an impact across the Atlantic, even with a particularly inert U.S. Congress. The Critical Medical Infrastructure Right-to-Repair Act of 2020 marked the first time in U.S. history that a right-to-repair bill was proposed on the national level. Since then, more than half the states in the union have introduced right-to-repair bills that call for equal access to things such as replacement parts, training manuals, and tools. While these bills are often limited to a specific industry—targeting electronics, appliances, automobiles, farming, or medical equipment—the passage of just one could have ripple effects across the nation. In November 2020, Massachusetts passed a resolution to further empower its 2012 automotive right-to-repair law—the first and only right-to-repair law on the books in the United States. The resolution expands the data and diagnostic information automakers are required to provide, thus enabling third-party repairs. Despite its limited scope, the 2012 law led to a national standard for automakers, and the recent resolution is expected to have similar effects.

Also noteworthy is the U.S. Department of Energy's launch of the Remade Institute in 2019, led by Nabil Nasr, an Egyptian-American engineer who had previously led the authorship of a report for the United Nations on the "Remanufacturing Revolution." The Remade Institute focuses on accelerating the circular economy through technological solutions and is grounded in five key areas of concentration (nodes): systems analysis and integration, design for "Re-X," manufacturing materials optimization, remanufacturing and end-of-life (EOL) reuse, and recycling and recovery. The growth of aluminum recycling transformed the industry in remarkable ways. In 1960, recycled aluminum accounted for 18 percent of America's total aluminum supply. Over the next forty-five years production of recycled aluminum rose by almost 746 percent; during those same forty-five years the total U.S. aluminum metal supply increased by 300 percent. The theoretical minimum energy required to produce secondary aluminum at 960°C is 0.39 kWh per kilogram; on a theoretical and practical basis, the energy required for secondary aluminum is less than 6.5 percent of the energy required to produce the primary metal.[13]

In late 2021, the European Commission has outlined plans to establish a new "right to repair" for consumers. Currently, EU consumers have a right to have faulty products repaired, but only when a defect is present at the time of delivery and becomes apparent within the legal warranty or guarantee period, which in most EU member states is two years. The European Commission launched a consultation in January 2022 on whether it should establish a consumer right to repair for situations not covered by the current legal warranty period.

The proposals by European Commission include:

- Creating a new right to repair for defects caused by wear and tear or mishandling of the product if this arises within a set period—potentially two years. This may only be applicable to certain categories, but the European Union has suggested that consumer products and electronics could be among these.

- Amending the Sale of Goods Directive to ensure that repair, rather than replacement, is the primary remedy available to consumers. Replacement of a defective product would only be available if repair is not possible or more costly than replacement.

- Restarting the legal warranty period for products that have been repaired, which means the consumer would have an additional warranty period of a minimum of two years after the product is repaired.

- Providing a longer legal warranty period to allow consumers to claim from sellers for repair or replacement of a defective product.

- Extending the legal warranty period for secondhand and refurbished products to equal that of new products. Currently, the parties can agree to a shorter liability period of not less than one year for secondhand products.

Despite these positive measures to improve resource usage efficiency and mitigate the wanton consumption of metals, we must not forget that there will still be massive material needs for a "green transition" in our economy. Moving from fossil fuels to renewable technologies will itself produce massive demand, which has now been widely studied and documented and has led to an entire field of "critical metals" research within economic geology.[14] Let us just consider the small microcosm of electronic devices as an example for how metal demand will continue to rise. The demand of iron, aluminum, and copper to produce new electronics in 2019 was approximately 39 megatons. Even in an ideal scenario in which all the iron, copper, and aluminum resulting from e-waste (25 megatons) is recycled, the world would still require approximately 14 megatons of iron, aluminum, and copper from primary resources to manufacture new electronics (11.6

Mt, 1.4 Mt, and 0.8 Mt, respectively). Therefore, any efforts at improving circular material flows will also need to be accompanied by the continued extraction of new geological resources. There is nevertheless massive opportunity for improvement in recycling of high-throughput products, such as beverage cans. The most recent data from the U.S. Environmental Protection Agency based on 2018 surveys suggests that only around 50 percent of aluminum cans are recycled (figure 7.5).

There are structural barriers to recycling a can versus a lead acid battery. With batteries, there is an institutional mechanism for retrieval, since most of the replacements are done at a workshop, which is heavily regulated for disposal streams. There is also a good economic incentive to retrieve the metals from the battery. With aluminum cans, that incentive may be missing. A five-cent deposit recovery helped initially to raise recycling rates in some communities, but this has plateaued, since the

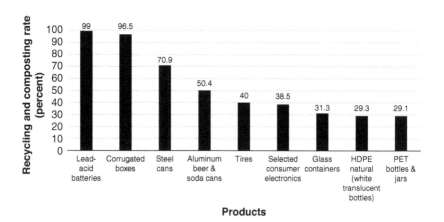

FIGURE 7.5 Comparative recycling rates of fast-throughput products.

Source: Adapted from U.S. Environmental Protection Agency, "Recycling Economic Information Report," 2020, https://www.epa.gov/sites/default/files/2021-01/documents/2018_ff_fact_sheet_dec_2020_fnl_508.pdf.

willingness to invest effort within the broader population is limited. Focusing on raising the recovery rates of cans for recycling through sorting of landfill waste or other engineered mechanisms of retrieval deserves prioritization, as it could be a win-win outcome. If we achieve a 90 percent recycling rate, we could save around 610,000 tons of aluminum—an amount equivalent to the annual production of at least three major American smelters.[15] Even research commissioned by the industry-led group International Aluminum using 2019 data noted that more than double the amount of aluminum is mined (around 77 million tons) than is in the circular-economy flow (around 32 million tons), and more than 24 million tons is annually lost as waste.[16]

The promise of a circular economy rescuing us from resource scarcity has potential for realization, but only if we can keep a systems perspective on material demand and consumption flows. In 2011, the UN International Resource Panel, of which I am now a member, estimated that less than one-third of some sixty metals studied have an end-of-life recycling rate above 50 percent. Indeed, thirty-four elements are below 1 percent recycling, yet many of them are crucial to clean technologies.[17] This number has improved marginally but certainly not kept pace with demand.

While aluminum foil is fully recyclable, the amount of foil that has been recycled has been very limited. Its thin cross-section and high surface area tends to rapidly oxide into dross, and product recovery was poor. Recycling of the combined plastic/foil aseptic packaging especially has been an issue. However, a group of manufacturers involving Alcoa, Tetra Pak, Klabin, and TSL Ambiental announced a solution to the recycling problem: use of a high-temperature plasma jet to heat the plastic and aluminum mixture. Under normal conditions the plastic is converted to paraffin, and the aluminum can be recovered in the form of a high-purity ingot that can then be used to manufacture new foil. The emissions of pollutants during the recycling of materials are reported to be minimal, and energy efficiency is close to 90 percent.

There is still a lot of potential for greater recycling but also limits to what we can recycle, given the durability dilemma for certain infrastructure investment. In this context, we also need a more expansive view of renewal—not just of the materials but of the ecosystems from which they were extracted. Minerals come from the earth, and through industrial processes we must alter ecologies to extract them. What happens to those altered landscapes and seascapes in terms of their other life-sustaining functions is the ultimate aspect of renewal we must aspire for. If properly managed, the restoration process of the land and the seas following extraction could also open opportunities for a different kind of social renewal, which we will explore in the next chapter.

8

RESTORATION AND RENEWAL OF MINERAL FRONTIERS

The graves like the yam vines, the coca trees, the houses, would soon disappear. The bones of their ancestors would be ground up in the red dirt, shipped off to another country to be immortalized in shiny aluminum saucepans and flying airplanes

—Olive Senior, "Boxed In"

Twentieth-century Caribbean literature tells many tales of transformed landscapes: forests made way for sugarcane and pineapple plantations, fishing communities gave way to privatized beach resorts, natural valleys hosted massive telescopes, and of course vast tracts of agricultural lands became bauxite mining projects. There is a yearning for a restoration of the landscape and a mourning that the extractive process often took with it many lifestyles linked with the land. The Jamaican Canadian author Olive Senior depicts this tale of loss in her poignant short story "Boxed In," in which a retired farmer reluctantly sells his property to a bauxite company in the 1960s. He languishes in front of a newly introduced television set in his final days, reminiscing about the magical attributes of his traditional taboos versus those of the technology before his eyes. Like the animated character displayed by the metallic television box, he sees his life "boxed in" by bauxite. He thinks about how his fellow Jamaicans were beguiled into selling their arable farmland as bauxite "became talismanic, as simple poor farmers . . . found their red soil turning to gold . . . tales of instant wealth, of men lighting their cigars with hundred-dollar bills, as they had done in the heady days of the building of the Panama Canal, of the Cuban sugar boom, of the banana barons."

Olive Senior's writings vividly capture the challenges that Jamaica and other densely populated islands faced when precious land was taken over for a variety of commercial uses, with promises of restoration that often did not materialize. Jamaica's saga of bauxite extraction connects us with the heyday of Caribbean aluminum ventures, when Alcoa ran cruises that highlighted the coexistence of tourism and the Caribbean. The posters showed idyllic vistas of tropical isles with exotic foliage and smiling locals in exotic attire ready to welcome American travelers aboard ships like the *Aluminum Clipper*. However, the story of bauxite in Jamaica is far more rancorous regarding qualitative impacts on the land and its ability to sustain an agrarian economy.

All commodities may well be subject to what are termed "Kondratieff Waves" of "boom and bust" economic cycles. These "waves" can last around forty to sixty years on average and were named after the Russian economist Nikolai Kondratieff, who theorized such tendencies in capitalist systems in the 1920s. By linking the macroeconomic performance of the United States, England, France, and Germany between 1790 and 1920 to the movement of commodity prices and production, Kondratieff explained these cycles in terms of technological change, geopolitics, and the price of gold. His theory suggested that capitalist nations were not doomed to failure; for this Stalin had him executed by a firing squad in 1938. A few years later, Joseph

Schumpeter published his magnum opus *Capitalism, Socialism, and Democracy*, in which he adapted Kondratieff's theory to his view of radical innovations and "forces of creative destruction." Each Kondratieff cycle is hypothesized to have four seasons: Spring (boom), Summer (expansion), Autumn (stagnation), and Winter. An appreciation for such cyclicality can help quell the populist-based nationalism that often takes hold in times of crises.[1]

As I was finishing the research for this book, I had an opportunity to interview a Jamaican American filmmaker, Esther Figueroa, who has dedicated her career to documenting social justice concerns in the Caribbean. Her documentary, which was released just before the COVID-19 pandemic, provides a critical environmental history of aluminum. Titled *Fly Me to the Moon*, the film features footage from the heyday of aluminum-product advertising and connects it to the struggle for land in Jamaica. During the postwar era, while Jamaica was still part of the British Empire, Marshall Plan funds were used to subsidize aluminum investment in the country. Colonial cycles of labor demand and struggles for patterns of migration for survival led many farmers to sell their lands to the bauxite companies. The farmers moved to other regions, which were often less arable. In an initial draft of this chapter, I errantly assumed that the land sales had simply been driven by the "allure of development," but as Dr. Figueroa corrected me in her review of the draft, the causes of land exchange for bauxite were highly complex and also linked to the Windrush Generation of migrants being attracted to the metropole for menial labor and postwar reconstruction in Britain. Meanwhile, the land of their ancestors, which they had often sold for a song, was extracted for mineral wealth and largely left in unusable condition for any future resettlement prospects.

When the land was eventually "restored," its decreased fertility could not sustain the same levels of food production. Original footage of farmers on their properties showed unkempt foliage on erstwhile fruit orchards that the government and the aluminum companies had claimed to have restored. The reality is that land where bauxite is mined is already challenged in terms of arability, and restoration after mining is highly difficult to achieve. Indeed, the original discovery of bauxite ore in Jamaica is credited to a farmer named Alfred D'Costa, who sent samples of his cattle farm's soil to test if it was suitable for particular grass varieties. As soon as the War Office in London discovered that high-grade bauxite was to be found, the British government declared all bauxite deposits to be the property of the Crown. Esther Figueroa also pointed out that the same story of Alfred D'Costa was also used by the bauxite establishment, with a twist. Industry claimed that he sent the soil for testing because it was not fertile and wanted to know how he could meet the food-production drive during the war with such low-quality soil, the subtext being that agriculture and bauxite were not competing for land. Extensive agronomic fieldwork has clearly shown that productive agricultural land was indeed acquired for bauxite production.

Island communities are rightly covetous of their scarce land and water resources, and when minerals compete with agriculture or forestry for arable acreage, friction and acrimony are inevitable. Mining has led to civil strife on numerous islands, most notably Bougainville and New Caledonia, in the South Pacific. In both cases, tension over the spoils of minerals development led to armed conflict and has also caused calls for independence. In the case of Jamaica, the conflict was somewhat contained, but the film footage captured by Dr. Figueroa of a consultation meeting between the farmers and a government minister documented a seething anger and resentment over the loss of land. In the postscreening interview, Dr. Figueroa situated the sense of plunder that Jamaicans have felt going back to the pirate era, when Port Royal was one of the largest cities in the Caribbean. This tiny peninsula near Kingston hosted ships of "privateers" who were commissioned by the British to attack their rival Spaniard colonialists. This was an age where resource wealth blinded the morals of empires and made commodities out of humans.

The pernicious history of the slave trade was also linked to land cultivation and access. Even though there is no connection between bauxite and slave labor, there was a post-traumatic connection for many farmers. After emancipation, many former slaves had been able to own small tracts of property—rather than being claimed as property themselves. When they were then cornered into selling their land for bauxite development, it added a degree of insult to injury. In most cases, the land was sold with the promise of rehabilitation. Yet, as Dr. Figueroa documented through footage across the island, the arability of the rehabilitated land was highly limited. In most cases, erstwhile farmers had to transition to other livelihoods by moving to urban areas. This legacy of dashed hopes of land restoration has led to a massive opposition to any expansion of mining in the central-western highlands of the island, in what is called "Cockpit Country." The usual criticism of mining from environmentalists hinges on the "nonrenewability" of the resource on human timescales. As argued earlier in this book, this is less of an issue chemically, as metals are themselves renewable within a circular economy if we can design products and invest in sustainable sources of energy. However, the degradation of land and water from the extraction process is a far more consequential question for livelihoods and for managing the impact of metal supply chains.

In early 2022, the Jamaican government authorized the mining of over 1,300 hectares of land in St. Ann for Noranda Bauxite, despite public protest by indigenous Maroon landowners. At the same time, the company also completed a reservoir in St. Ann dedicated to combating the ever-present bauxite dust problem by wetting the roads over which transport trucks travel. Jamaica's transport and mining minister Robert Montague inaugurated the new reservoir, which holds about 17 million gallons of water, on December 21. The company noted that the reservoir was built in response to continued complaints by the community about bauxite ore dust. The new reservoir will allow Noranda to continue to wet the roads in the area; it distributes up to 70 thousand gallons of water each day. The cost of the reservoir totaled US$800,000—a mere fraction of the overall revenue and profit base of the company.[2]

A COCKPIT VIEW OF HISPANIOLA

A few hundred miles east of Jamaica, the island of Hispaniola provides geographers with a rare natural experiment of how history, landscape, and industrial development can produce highly divergent trajectories to nationhood. The second-largest island in the Caribbean hosts two countries that are often presented as drastic contrasts. While attending a seminar at a NASA event in Washington some twenty years ago, I remember being shown a satellite image of the island.[3] The speaker asked everyone if we could tell where the border between Haiti and the Dominican Republic was. Everyone nodded in unison: you could see a dramatic difference in land cover from the satellite image, coincident with where the border lay. The Haitian side showed far less vegetation. The Pulitzer Prize–winning author and ecologist

Jared Diamond dedicated a chapter of his book *Collapse* to why Dominicans protected their forests while Haitians degraded theirs. The border photo appeared in Al Gore's 2006 documentary *An Inconvenient Truth*.

Subsequently there have been revisionist accounts on influential media outlets like *Vice*, claiming that "one of the most repeated facts about Haiti is a lie,"[4] the underlying accusation being that the deforestation difference is overplayed and a reflection of white prejudice against the poor Haitians. However, there is now ample evidence from the field to show that there is indeed distinctly higher deforestation on the Haitian side. A study in 2018 published in the *Proceedings of the National Academy of Sciences*, using multiple methods and focusing on primary forests, calculated that such high-biodiversity regions in Haiti declined from 4.4 percent of total land area in 1988 to 0.32 percent in 2016.[5] Nevertheless, in a remarkable show of adaptive capacity, Haitians have managed to find a way for the carefully managed woodlots known as *rakbwa*, in which trees are grown, culled, and sold for construction or charcoal, to play a role in sustaining the country's tree cover.[6] The clear difference between the two countries on Hispaniola is an index of their diverging histories.

The eastern side of Hispaniola was colonized by the Spanish and the western side by the French, leading to linguistic differences but also a fairly different political culture, though they both shared a tragic common heritage of slavery in plantations across the island during the sixteenth and seventeenth centuries. Following the French Revolution, slavery was abolished by France in 1793. However, avarice and prejudicial suspicion of the emancipated slaves continued in the Caribbean, prompting a threat of the reestablishment of bondage in 1802. This led to the Haitian Revolution, toppling the French colonial powers completely, the first such triumph of oppressed slaves against a European power. As the first independent nation to emerge from resettled African slaves, in 1804 Haiti held much promise at its inception. In fact, the Haitians ended up controlling all of Hispaniola from 1822 to 1844, ending slavery on the Spanish side of the island as well. However, the Dominican settlers resented being ruled by the Haitians, and after several years of conflict both countries agreed to recognize each other's respective sovereignty in 1856.

The nascent Haitian state was marginalized by its neighbors, particularly the United States. The American government did not even recognize Haiti until 1861, for fear that the slave rebellion against the French colonial masters that created the country might inspire a similar uprising on Southern plantations. However, as the United States and the rest of the hemisphere finally embraced civil rights in the twentieth century, Haiti became a major locus of interest for investment and development. During the Cold War, Western nations supported despotic elites to serve their own interests in the region, but in recent years, there has been a more genuine commitment to helping Haiti climb its way out of poverty. Haiti is now the poorest country in the Western Hemisphere and has been struck by natural misfortunes and endured malevolent foreign intervention for decades.

Both Haiti and the Dominican Republic have mineral and other natural resources. When the French lost their prize colony in 1804, they levied an indemnity on the new Haitian government as punishment. To pay this debt, Haitians began exporting mahogany to France; by 1842, they were sending 4 million cubic feet of it overseas every year. Deforestation of the high-value primary hardwoods largely occurred in the nineteenth century, when the Haitian government turned over its mahogany forests to foreign forestry companies.

Haiti and the Dominican Republic both lie in the path of hurricanes, but the densely populated and mountainous terrain of Haiti is more vulnerable to landslides, which have been made more frequent by anthropogenic deforestation. The 2010 earthquake that struck Haiti's capital region was the worst calamity in the country's recorded history. Over two hundred thousand people were killed and over 1.6 million displaced in a country of just over 10 million. The aid effort was monumental but perhaps misguided. Cholera epidemics broke out after mismanaged or negligent United Nations peacekeeping efforts, and much of the aid funds did not reach Haitian organizations. A recent estimate indicates that over 9 billion dollars in aid came into Haiti since the earthquake but that 89 percent of the funds went to non-Haitian organizations. Thus restoration efforts can themselves generate a "disaster economy" that needs to be monitored to ensure that greater benefits flow to those in need.

Against this backdrop of despair and devastation, Haiti is ready for some better news. The discovery of mineral deposits with an estimated worth at current prices of around $20 billion has brought cautious optimism to some quarters of the country. In 2014, I was invited to Haiti to speak at a conference organized by the World Bank to consider a revival of the country's mineral industry. There was much anxiety and anticipation over the prospect of minerals providing much-needed domestic revenue generation, ideally decreasing the country's perpetual dependence on foreign aid. Haiti's only major experience with modern mining was a bauxite mine run by the Reynolds Corporation in the Miragoane region, dating back to 1941. From then up to 1982, the country exported 13.3 million tons of bauxite from Haiti to Reynolds's alumina refinery in Corpus Christi, Texas. Haitian bauxite accounted for almost one-fifth of Reynolds's bauxite acquisition in that period, and Reynolds was given access to 150,000 hectares, expelling thousands of Haitian families. The Haitian economist Fred Doura describes the extractive mining in Haiti as "a typical example of an 'enclave' industry under foreign domination where two North American transnationals exploited the minerals bauxite and copper. . . . The impacts was practically null on the economy."[7]

Currently, the largest private service-sector investment in Haiti thus far is the Royal Caribbean's Labadee Cruise Ship resort, which I visited a few years earlier. There are important ecological factors that should be considered before mining proceeds, and funding resources are available to Haiti through mechanisms such as the Critical Ecosystems Fund, which has supported some grants in northern parts of Haiti where mining is planned. Minerals did not deliver the wealth that Hispaniola had yearned for; instead, service sectors such as tourism have done far more for the economy of both countries. No doubt the COVID-19 pandemic showed the vulnerabilities of tourism in terms of revenue reliability, as have numerous hurricanes and natural disasters over the years. However, for highly populated areas where land is scarce, conservation efforts become a priority.

In 2017, both Haiti and the Dominican Republic agreed under the auspices of the UN Environment Programme to work jointly on putting to rest sterile narratives of the grass being greener on the other side. Through joint land-use management and restoration practices, they would ensure that the island's land was rehabilitated. Future land disruptions, such as from mining operations, could also benefit from the lessons learned from this joint eco-regional activity. Such a transformation of the narrative is encouraging.[8] However, the hunger for critical metals for the low-carbon transition, including aluminum, will drive further interest in mining on Hispaniola. The lessons in restorative land degradation will be put to the test on the island once more, as bauxite mining gains momentum in the Dominican Republic and the port of Cabo Rojo (Red Cape) and the nearby Jaragua National Park contend with the persistent challenge of "red mud."

MUDDIED REDEMPTION

One of the major byproducts of the Bayer process for producing alumina from bauxite is the iron-rich slimy waste colloquially called "red mud." Over 95 percent of the alumina produced globally is through the Bayer process; for every ton of alumina produced, approximately 1 to 1.5 tons of red mud are also produced. Annual production of alumina in 2020 was over 133 million tons, resulting in the generation of over 175 million tons of red mud. While there is no good tracking system for red mud, an estimate published in the journal *Hydrometallurgy* in 2011 suggested that 2.7 billion tons of red mud are stored worldwide—often in precarious impoundments. This figure was growing on average by 120 million tons per year at that time.[9]

Red mud is a result of bauxite processing, although there have also been efforts made to extract aluminum from other sources of the metal. Russia invested enormous resources to extract aluminum instead from nepheline—a form of aluminum silicate mineral often found coincident with apatite (a calcium phosphate mineral). Apatite is used in fertilizer manufacturing, and Russia saw an opportunity to efficiently mine both aluminum and phosphates from the same deposits in the Kirovsk region. The apatite processing often left tailings with a nepheline content as high as 24 percent. As early as June 1955, a decision was made to build an alumina plant at Achinsk in Siberia to process nepheline ores. Further plans were constructed in Razdan, Armenia, and Kirovabad (now Ganja), Azerbaijan, in the 1960s. However, many of these efforts were not economically viable without state subsidies motivated by localized resource-security arguments called *mestnichestvo*. The historian Stephen Fortescue notes that these efforts during the Khrushchev era stressed "regional development and self-sufficiency, pushed by local political and scientific elites." Even though such production was ecologically more appealing than some forms of bauxite production, the competing economies of scale for bauxite production elsewhere made innovations in these processes less viable. More recently, as the ecological cost of red mud has moved further to the forefront, there is renewed business interest in further innovating to retrieve aluminum from nepheline.[10]

In October 2010, heavy rain caused a red mud dam to collapse at a containment facility in Ajka in western Hungary. The incident became one of the most widely publicized environmental disasters in Europe. Around a million cubic meters of sludge were discharged into the Torna-Marcal river system, the Hungarian countryside, and ultimately into the Danube. This was the largest incident of such massive discharge of red mud into the environment. Although the immediate impact of the mud was devastating, the long-term impact has since been calculated to be far less than expected. Indeed, the mining of bauxite itself and inadequate land remediation in bauxite mining sites demonstrate far more long-term impacts. This may be credited to the Hungarian government's timely remediation efforts, which cost over €127 million. Much of the funds were spent on dredging mud out of rivers and removing it from the floodplain. Sites near the source of the spill were also dosed with acid and, further away, with gypsum to reduce the dangerous levels of alkalinity. Research conducted by the University of Hull revealed that any geochemical signal from the red mud had almost completely disappeared within three years.[11] The fine-grained nature of the mud made it easier for any residuals to get quickly washed downstream and into the Danube, where they were diluted to an innocuous level. Air-quality impacts of the red mud dust on human health also showed less impact than expected, thanks to the chemical and physical properties of the mud grains.[12]

Nevertheless, the European Union was traumatized by this incident, and in 2015 a major initiative was launched in Europe to find some ways of utilizing red mud for products to mitigate the risk of its buildup in dams. Fifteen doctoral students were recruited as part the European Training Network (ETN) for Zero-Waste Valorisation of Bauxite Residue. Recovery of iron, aluminum, titanium, scandium, and rare-earth elements was prioritized, and the remaining residue's potential for use in building materials was explored.[13] Subsequently, a European Innovation Partnership has been formed to consider a circular-economy approach to red mud titled BRAVO (Bauxite Residue and Aluminum Valorisation Operations). This project has endeavored to bring together industry with researchers and stakeholders to investigate mechanisms for commercializing the retrieved critical raw materials from red mud. In 2018, further funds were allocated for a project titled RemovAl, which is encouraging entrepreneurs to consider innovations that could lead to the efficient use of red mud in a range of infrastructure efforts.

Through the EU research, a "red mud house" made entirely out of materials from bauxite residue is being prototyped. The twenty-five-square-meter demonstration house will be built in a housing settlement next to an alumina plant in the Aspra Spitia area of Greece, which means "white houses." Professor Pontikes from the Catholic University of Leuven in Belgium is managing this effort and notes that "the final building elements will actually have a wide range of colors." The red mud allows for a design with interlocking building blocks that fit into one another so that the residence can be dismantled and deconstructed instead of demolished. Such a Lego-style approach would also make reuse and reassembly easier and be in greater congruence with circular-economy principles than a concrete/mortar construction.[14]

Despite the pandemic, in November 2020, the ReActiv: Industrial Residue Activation for Sustainable Cement Production research project was launched by one of the world's largest cement companies, Holcim, in cooperation with twenty partners across twelve European countries. The ReActiv project will create a "novel sustainable symbiotic value chain, linking the by-product of the alumina production industry and the cement production industry." The project aims to link the alumina and cement value chains through new technologies. The latter will modify the properties of the industrial residue, transforming it into a reactive material (with pozzolanic or hydraulic activity) suitable for low-CO_2-footprint cement products. In this manner, ReActiv proposes a win-win scenario for both industrial sectors (reducing wastes and CO_2 emissions, respectively). Figure 8.1 shows a summary of the uses of red mud that are now being developed to make this lamentable waste byproduct more valuable.

The efforts to deal with the red mud problem have also had positive spillover effects, with researchers considering a range of other tailings facilities and waste-rock dumps as sources of useful ore. Rio Tinto Corporation has commenced production of battery-grade lithium from waste rock at a lithium demonstration plant at a mine site in California. The plant will aim to have an initial capacity of at least five thousand tons per year, or enough to make batteries for approximately seventy thousand

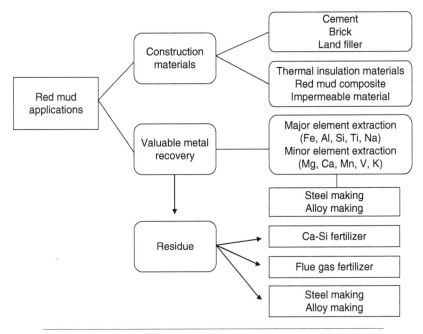

FIGURE 8.1 Various uses of red mud being developed.

Source: Adapted from S. Samal, "Utilization of Red Mud as a Source for
Metal Ions—a Review," *Materials* 14, no. 9 (2021): 2211.

electric vehicles. The Canadian government has also launched a
program to look for mineral deposits in abandoned tailings sites
across the country. Researchers at the University of Queensland
in partnership with the University of Exeter and the University
of Geneva have established a Global Tailings Research Consor-
tium to further such efforts on multiple fronts. The extraction of
sand from tailings worldwide could reduce the pressure on coastal
systems, where sand is often mined for construction purposes.[15]
In April 2022, Dr. Longbin Huang from this program announced
that they had successfully delivered the first field-operable and
scalable technology to ecoengineer bauxite residues into fertile

soil within a defined timeframe, for soilless rehabilitation of plant communities.[16] Such technical solutions are heartening, but they must also be coupled with a healing process for communities already affected by past extractive episodes.

THE CHALLENGE OF
SOCIAL RESTORATION

Since the early days of the aluminum industry, communities along the St. Lawrence Seaway in the northeastern border region between the United States and Canada have been at the frontlines of impact—for better and for worse. Endowed with massive hydropower and maritime access, industrial investment gained momentum on both sides of the border, often with little regard for the indigenous inhabitants of this land. During the decade when I was a professor at the University of Vermont, from 2002 to 2012, I would drive through this region on my way to Toronto to visit my sister. My family and I would stop for gas at New York State's tribal filling stations, which were cheaper, thanks to different taxation rates from the rest of the state. Smokestacks and decaying industrial buildings could be seen in the distance as we crossed the bridge across the seaway. When an opportunity arose to suggest a topic for a doctoral dissertation to my students, I would often note this region as ripe for research on environmental injustice.

After several years of trying to garner interest in a case study set in that region, a midcareer social service professional named Kim McRae expressed an interest in taking on this topic. Kim had a background in social justice advocacy and wanted to connect that with environmental concerns. As an African American woman living in a welcoming and green but very racially homogenous state like Vermont, she was keen to understand how other ethnic minorities had reconciled environmental conservation with cultural integrity. Growing up in the metal and concrete of the Bronx, she thought of nature as a suburban phenomenon quite distant from her own life. Every summer she would travel to an interracial community called Camp Unity, where her parents had first met as teenagers. Her mother was from a conservative Jewish family, and her father was a Black southerner. Troubled by the inequality that rapid industrialization had brought to their lives, they gravitated toward Camp Unity to seek healing but also by their impulse for restorative justice. There is ample research showing that being situated in natural environments, particularly amid foliage, can have a restorative and healing impact on communities. Many years later, when Kim decided to do her doctorate, she sought a topic that would connect social justice with the ecological restoration of industrial landscapes.

The plight of the Akwesasne community, within a few hours' drive from the camp where her parents had fallen in love, became a personally and professionally relevant case study for her dissertation. Nevertheless, despite her diverse background and keen awareness of cross-cultural sensitivities, it took Kim several years to build trust with the community of Akwesasne to get access for ethnographic research. However, she ultimately prevailed and produced an important dissertation on how PCB contamination from various automotive plants had affected the community.[17] The automotive plants were located in the region partly because of the proximity of aluminum plants. The region had become an industrial cluster, which was economically efficient but had a multiplier effect in terms of damage to the community. How this community has tried to heal and reconcile with its industrial legacy is a keystone case study of endurance.

The industrial legacy of this region has made the community intensely sensitive to environmental-management concerns. I interviewed Abraham Francis, the environmental manager for the Mohawk Council of Akwesasne, to gain further insight on how the community was moving toward ecological and social restoration. Abraham had undertaken a master's degree in natural resources and a bachelors in biology at Cornell University; he shared his master's thesis with me. This is an impressive document on indigenous forest stewardship in his region, and in reading it I gained a better understanding of the community's efforts at social restoration while navigating the vagaries of colonial demarcations. Politically, Akwesasne is a single community that has three governments: in the northern portion ("Canadian") is the Mohawk Council of Akwesasne (MCA); in the southern portion ("American") is the Saint Regis Mohawk Tribe (SRMT); and within the Haudenosaunee Confederacy is the holder of the fire for the Kanienkehaka, the Mohawk Nation Council of Chiefs (MNCC). Akwesasne resides at the intersection of two Canadian provinces, Ontario and Quebec, and an American state, New York, which leads to complicated issues of jurisdiction and funding.

Abraham states in his thesis that

> through a recounting of Akwesasne's political history, the complexity of relationships between *Akwesasronon* (citizens of Akwesasne) with each other, themselves, settler-colonial states, and the land becomes apparent. It is thanks to foundational research by *Akwesasronon* and their collaborators that the cultural dimension of the environmental violence in and around Akwesasne were acknowledged and engaged. This shows the resiliency and persistence of the community to not be silenced and desire to heal historical wounds left by environmental violence.

Through the persistence of individual activists from the community, they were eventually able to receive $8,387,898 in repara-
tions for the damages done to Akwesasronon culture. The funds
were to be utilized for tribal cultural restoration projects.[18] The
community does not consider the U.S.-Canadian border as consequential to their lives and even during the COVID-19 pandemic, when the border was closed, there was movement across
tribal territories. As Abraham noted, legal victories aside, social
restoration comes to "incorporating cultural values as part of
their organizational culture and structure."

Brenda LaFrance, an Akwesasne elder, spoke with dignified
composure about how the industrial legacy in her community
had affected her. "We were listened to but not heard," she said
of her encounters with industry and government alike. Brenda's husband worked for Alcoa for much of his life. He enjoyed
his work but ended his career with Alzheimer's disease, which
his family ascribes to aluminum poisoning. Reynolds also had
a plant in the same vicinity, as did General Motors, and these
industries were all upstream of the Akwesasne community. The
hydropower bonanza brought these companies to the region. I
asked Brenda what she envisaged the future would have been if
industry had not come to her lands. She reminisced about how
fishing; trapping muskrats and beavers; and hunting deer provided ample meat and that owning dairy cattle was the norm in
households. There was abundant fish for home use and also for
commercial sales. The St. Lawrence had fantastic sturgeon, and
the eggs were shipped to New York City for caviar. There was a
sturgeon-smoking facility on an island in the river that the
community managed. The bullhead was another common, edi-
ble, and delicious fish, but as a bottom feeder, it became most
poisoned by the pollution.

The advent of aluminum and its adjoining industries was a disruptive event in the lives of the Akwesasne. But Brenda was not opposed to industry—only to how it was practiced. Finding pathways for the coexistence of industry and subsistence lifestyles requires us to consider how metals enter the biotic system and the roles they play in our diet and our bodies. We are what we eat—this is more than a cliché, and it has ramifications not just for human health but the overarching functionality of a society. Humanity has developed networks of influence, caring, and trust through food, and any disruption to diet can have an effect on cultural connectivity. After our call, Brenda reflected further and two days later sent me a moving email in which she laid out her feelings about the effect of industrial pollutants on her community: "Multigenerational relationships were severely impacted by the contamination found in fish and wildlife, and the soils in Akwesasne. Our people moved away from the rivers and gardening declined rapidly as they were seen as poisonous and not to be consumed."[19]

Nevertheless, human systems can be restored, just as ecosystems regenerate from a forest fire, so long as a kernel of knowledge is there for resurrection. The elders of the community tenaciously preserved their language and vestiges of oral tradition about methods of hunting and fishing. Once the pollution abated and a cleanup of the waterways commenced through a series of protracted litigations and regulatory enforcement, the biotic system bounced back. The St. Regis community is again selling fish in restaurants, although there is still a restriction of only one or two fish per month to moderate metal loads in diets.

Many of the Mohawk youth started to attend college in the 1990s and studied environmental science, which helped bring knowledge of modern analytical techniques to the reservation. As a transboundary region, bordering Canada, they were providing adequate livelihoods for the area. The tribe has a sustainable agriculture program, which gives the tribe hope.

The Akwesasne community members often refer to their Mohawk Thanksgiving Prayer[20] as a source of inspiration for their approach to restoration. As an oral history passed from one generation to the next, the prayer highlights the tribe's commitment to environmental responsibility. Many of the natural resources in the region are noted, followed by an explanation of how to conserve this legacy by paying attention to even the most diminutive of creatures. There is frequent mention of grasses and plants and the services they provide to the community. This is a recurring theme across indigenous traditions. Social restoration and the plant kingdom have a special connection, one that has now also become apparent at more technical levels: plants provide ecological healing while also delivering valuable resources.

THE BOTANICALS OF
NATURAL HEALING

We often do not connect metals with vegetation, but all botanicals have some metals in their biology. Finding a way to harness metals from plants or to use their unique properties to absorb metals from the soil has gained considerable attention in recent years, alongside the role of microbes in metallic metabolism. In a landmark paper published in 1997, Rufus L. Chaney, a botanist at the U.S. Department of Agriculture, proposed the use of plants in the remediation of contaminated soils using a new technique called "phytoextraction."[21] His research involved plants that are called "hyperaccumulators," so called because they have the capacity to disproportionately tolerate and store metallic compounds in their sap. During a field visit to New Caledonia,

I encountered one of these plants, which was able to take up so much nickel from the rich lateritic mineral soil of the island that its sap was bright green.

Cheney's technique to use these plants instrumentally for soil remediation and also to extract some useful metal products involves (1) cultivating selected hyperaccumulator plants on a contaminated site and (2) removing the harvestable metal(loid)-enriched biomass to reduce the volume of plant material disposed of as hazardous waste. Phytoextraction can remove hazardous metal(loid)s from the soil in a cost-effective way and compares favorably with other available remediation techniques, such as physicochemical methods of decontamination. Phytomining—through subsequent metallurgical processes to recover valuable metal(loid) elements from the biomass—can create a profit from these metal(loid)s.

Elements such as cobalt, nickel, selenium, thallium, and some rare-earth elements (REEs) are considered as critical, given their limited availability and increasing demand. The high market price of these elements makes them ideal for phytomining. In the case of nickel and cobalt, phytoextraction can be applied to low-grade and agriculturally unproductive ultramafic soils, which naturally contain high concentrations of these elements and cover more than 3 percent of Earth's land area. Phytomining can also be applied to seleniferous soils, where the prevailing concentration of selenium is high. These soils cover exceptionally large areas in Australia, the United States, and other countries. Worldwide, abandoned mining waste left without sufficient remediation could be considered a raw material for phytomining. Metal(loids) without a profitable market price (for example, arsenic or cadmium) can also be phytoextracted from the waste to reduce toxicity, ultimately improving the soil

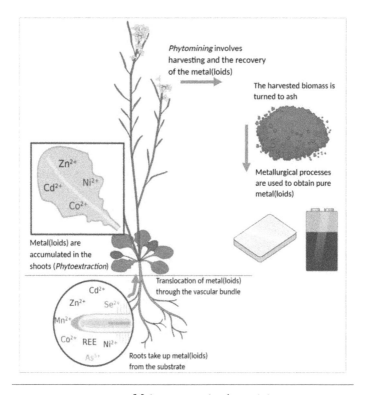

FIGURE 8.2 Main processes in phytomining.

Source: Figure created in Biorender.com and Mind the Graph platform
by University of Queensland, Sustainable Minerals Institute.

geochemistry and allowing the return of native plant species
and land remediation.

Cobalt is a critical commodity. Global demand for it is increasing, and the highest-grade ores occur in just one location: more
than 50 percent of the world's cobalt supply originates from the
Democratic Republic of the Congo, an area suffering considerable sociopolitical instability. Cobalt is used to build lithium-
ion batteries for electric cars and is most often a byproduct of

copper and nickel mining. Phytomining of cobalt from current and abandoned mines would offer an alternative approach to obtaining the element. For example, the large-scale laterite nickel-cobalt mines in Australia (Murrin Murrin), former cobalt mines in the Democratic Republic of the Congo (owned by the mining company Gécamines), and artisanal mining activities in Katanga all produce massive volumes of waste material with a suitable cobalt content. The hyperaccumulator plants *Haumaniastrum robertii* and *Berkheya coddii* are potentially good candidates for phytomining in these scenarios. The biomass yield of *H. robertii* is estimated to be up to 5 tons per hectare per year, containing cobalt on average at 5,000 mg/kg biomass, for an annual cobalt yield of 25 kilograms, which is worth US$1,350 (excluding production and processing costs).

REEs comprise seventeen metallic elements (fifteen lanthanides, plus yttrium and scandium) widely distributed throughout Earth's crust. Cerium, lanthanide, neodymium, and yttrium are the most abundant. Recently, these elements have been used in a myriad of applications, such as green technologies, medical instruments, and microfertilizers. The increasing demand for REEs has resulted in their limited future availability, and their potential future release to the environment poses a risk to numerous ecosystems. Phytoextraction can remove these elements from polluted soils, and phytomining can commercially produce high-value REEs.

The market price for REEs in their oxide forms depends on the element and its purity. For example, cerium and lanthanide have low prices (~US$5/kg), whereas terbium and dysprosium are currently valued at more than US$200/kg. *Dicranopteris linearis* is a hyperaccumulator fern that can concentrate REEs at up to 3,358 mg/kg biomass in its fronds; however, it contains 47.5 percent lanthanide and just 6 percent dysprosium. Even though REE yields of up to 300 kg per hectare have been estimated for *D. linearis* (based on 15 tons of biomass containing REEs at 2,000 mg/kg biomass), REE phytoextraction using this particular hyperaccumulator is not currently economically feasible given the low price of lanthanide and cerium.

The use of hyperaccumulator plants for phytoextraction and phytomining offers a series of benefits, such as the natural concentration of the desired elements and the exclusion of unwanted elements. The economic feasibility of phytomining, however, depends on the ability to recover the metal(loid)s of interest from the harvested biomass. In most cases, the harvested biomass is incinerated to ash to obtain "bio-ore," which greatly increases the metal(loid) concentration. Most work has focused on nickel recovery, in particular from *Odontarrhena muralis*, *Rinorea bengalensis*, and *Phyllanthus rufuschaneyi*, using either ashing or leaching of dry biomass followed by further refining to obtain pure metal, salts, or nickel catalysts. *O. muralis* contains nickel at 20,000 mg/kg biomass, translating to 32 wt% nickel in the ash. *P. rufuschaneyi* has extremely low concentrations (0.1 wt%) of unwanted contaminants such as iron, chromium, silica, and manganese. Recovery of REEs from *D. linearis* biomass has also been studied, including leaching processes and purification using ion-exchange resins or selective precipitation. To date, no work has been undertaken to recover cobalt or thallium from hyperaccumulator biomass.

Phytomining is a unique and relevant technology that pairs resource acquisition with environmental remediation and low-waste resources. Numerous abandoned mine waste and metal(loid)-enriched soil localities globally may be strong candidate sites for the installation of hyperaccumulator plants. Phytoextraction and phytomining have been trialed in experimental settings; however, they require testing at field scale to assess their broad-scale commercial potential.

DEMOGRAPHIC POWER AND UPSCALING INNOVATION

The world's two largest countries—China and India—have both become major aluminum powers in the past few decades. Although their political systems differ, they share a common cultural passion for science and engineering. The "demographic dividend" of young innovators from these two countries could be better harnessed to focus on solving some of the challenges posed by land restoration. China's mineral sector dominates the world at various stages of production and also in downstream manufacturing. The technocratic model of governance followed by the country has made mineral governance quite nimble but has also caused concern about the transparency of monitoring impacts. In 2012, the United States, the European Union, and Japan lodged a complaint with the World Trade Organization (WTO), arguing China's restrictions on REE exports were designed to provide Chinese industries with protected access to the minerals for use in domestic downstream industries. Ironically, China had claimed that the supply constraint had come about because the government had shut down many pollution-intensive mines to address environmental degradation. The plaintiffs' countries were thus in the uncomfortable position of asking China to open these dirty mines when they had previously criticized its environmental performance. In March 2014, the WTO ruled in favor of the complainants, asking China to remove its export tariffs and quotas or face repercussions. China had imposed three distinct types of REE export restrictions, including duties (taxes) on the export of the minerals, an export quota, and certain limitations on the enterprises permitted to export the materials. In April 2014, China appealed the decision; however, the WTO maintained their ruling.

In January 2019, less than a year before it became the epicenter of the COVID-19 pandemic, I was invited to visit the Chinese University of Geosciences in Wuhan. The sheer scale of research on improving the efficiency of mining at this institution is impressive. There are over thirty thousand students currently enrolled, and since its establishment in 1952 the university has graduated over three hundred thousand individuals. These graduates include former Chinese premier Wen Jiabao (2003–2013). Such a brain trust when channeled toward ecological solutions to the land-degradation challenges of mining could perhaps be a beacon of hope for win-win opportunities. The English version of the website for the university is emblazoned with the words "A Beautiful China and a Habitable Earth: Toward 2030." The choice of the year alludes to the year that the United Nations' seventeen Sustainable Development Goals and their associated 140-plus targets for human prosperity are aimed to reach fruition. Whether or not such institutions will be able to meet these challenges, it is essential to continue to galvanize such talent for finding solutions to these urgent degradation challenges, particularly in China, which controls so much of the world's mineral and metal supply chains.

The world's largest democracy—India—has been struggling to navigate a path between resource-linked development and land conservation. India's bauxite sector has come under particular scrutiny by celebrity writers such as Arundathi Roy because many of the deposits intersect with indigenous land claims, particularly in the Khondalite Mountains. The Birla family has controlled many of these deposits since the 1950s, when it formed a partnership with Kaiser Aluminum. G. D. Birla and Edgar Kaiser (a grandson of the legendary founder of the company, Henry Kaiser) formed a close camaraderie and established Hindalco—a company that also established several hydroelectric dams in the 1960s to support the smelting. Unlike the dams in Quebec, these impoundments ended up displacing more than 350,000 people and inundating valuable land, which made restoration prospects even more problematic. Tribal elders in the Kondistan region popularized the phrase "we are being flooded out with money" in their protest movements.[22] The Birla family has a distinguished history in India and its members were great supporters of Mahatma Gandhi, who spent the final four months of his life at the Birla residence and was assassinated by a Hindu extremist on their lawn. However, there was a stark contrast between the humanitarian impulses they proclaimed in politics and the perception of their industrial enterprise in the bauxite region. Such a "distance" between political power and impact perception remains a challenge in autocracies and democracies alike.

However, India's youthful scientists and engineers are beginning to engage with ecological planning for mining ventures so that the tragic land impacts of Kondistan are not repeated. A comprehensive anthology on *Innovations in Sustainable Mining* was assembled in 2021 by the Jawaharlal Nehru Aluminum Research Development and Design Centre in Nagpur in partnership with the National Institute for Occupational Health.[23] There were a range of well-researched innovations noted, including the use of red mud as a pollution adsorbent for a variety of chemical processes, as shown in figure 8.3.

The extraction of minerals from fly ash produced as a waste of mineral processing is another area where technological

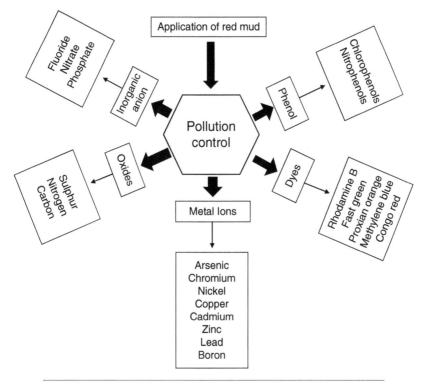

FIGURE 8.3 Applications of red mud as an adsorbent in pollution control.

Source: Adapted from S. Rai et al., "Utilization of Aluminum Industry Solid Waste
(Red Mud/Bauxite Residue) in Pollution Control," In *Innovations in Sustainable
Mining: Balancing Environment, Ecology, and Economy*, ed. K. Randive et al.
(Cham: Springer, 2021), 21–43.

innovations could play a vital role in improving the performance
of the sector. The annual world generation of fly ash is nearly
800 million tons, of which India alone produces more than 200
million tons. In India, around 43 percent of fly ash is utilized in
different sectors, including cement (major portion), leaving a
vast quantity unutilized, while in China and the United States
utilization rates are 70 percent and 50 percent. Fly ash contains

a range of valuable minerals, such as mullite, quartz, hematite, magnetite, alpha-alumina, calcium oxides, titanium oxides, and rare-earth elements, in a matrix of aluminosilicate glass. Nearly 50 million tons of alumina, the major material of value present in fly ash, is getting wasted because suitable technologies in India to recover it are unavailable. The disposal of such a huge quantity of fly ash also creates environmental problems. The greatest disadvantage of fly ash processing is its high silica content, because of which various conventional methods cannot be employed. Furthermore, the main alumina-bearing mineral is mullite, which is refractory in nature and difficult to solubilize.[24]

Approximately twenty kilograms of hazardous waste are produced for every ton of aluminum. Much of this waste is termed "spent pot lining" (SPL)—which is a mix of the carbon lining and the refractory lining of reaction vessels and contains high levels of leachable cyanides and fluorides, as well as components that form combustible gases (mainly methane). The current practice of the waste management of hazardous SPL is still often landfilling or incineration, which costs aluminum producers on average $250 per ton of SPL waste (around $300 million annually on a global level). The European Union Institute for Innovation and Technology on Raw Materials (EIT-raw materials) research program has developed and optimized the SPL-CYCLE project for wastes originating from two aluminum smelters. The new process consists of five main separation and purification stages, which involve dilution, filtration, crystallization, and flotation, resulting in the production of four products: (1) fluoride salts for aluminum production, (2) graphitized carbon for aluminum production, (3) aluminosilicates for refractory industry, and (4) manufactured aggregate for construction (supplementary cementitious materials, lightweight aggregates, geotechnical fill, bricks, concrete).

OCEANIC FRONTIERS

Apart from finding opportunities for a circular economy and existing mines' and waste reservoirs' efficiency, there may also be an opportunity to consider oceanic mineral reserves in the future. A key argument for such ventures is that many of the deleterious impacts of mining on land could be mitigated in deep waters, as there would not be the kind of property rights and competing land-use challenges in open waters. Given the rapid rise in demand for minerals and the declining ore reserves on land, attention is turning to the potential extraction of marine mineral deposits. While coastal marine mining for diamonds and for mineral sands has been undertaken for some decades, deep-sea mining is still in the initial stages of development. The UN Law of the Sea Convention (UNCLOS) established the International Seabed Authority in 1994 to develop a regulatory regime and issue licenses for mineral exploration in open waters beyond national jurisdictions. In some cases, the countries are small-island developing states, such as Nauru, which have established regulatory bodies and invested in exploration companies.

However, the environmental impact of oceanic mining remains widely contested,[25] and as the process of environmental regulation is formulated by the International Seabed Authority in coming years, attention will need to be paid to the following key issues:

- Sediment dislocation and plumes being generated by mining activity
- Impact of mining activity and noise on biodiversity
- Potential release of deep-sea carbon through extractive activity
- Impact of mining on fisheries and resultant livelihoods

Oceanic mineral extraction, however, does have technical advantages over terrestrial mining, as many of the ores are highly concentrated, and thus there would be reduced waste and water usage for processing, a lower carbon footprint for processing, and much less social impact on communities in terms of physical displacement and adverse effects in local livelihoods. Three main kinds of oceanic mineral deposit are being considered owing to their comparative ease of extraction and given their economic-ecological cost balance: polymetallic nodules, cobalt-rich crusts (which occur on some, but not all, seamounts), and seafloor massive sulfides (which are sulfide deposits from extinct hydrothermal vents). Much of the International Seabed Authority's interest is currently focused on polymetallic nodules in the Pacific Ocean.

Preliminary comparative analysis of the carbon footprint of terrestrial and oceanic mineral processing, as well as the waste-generation and water-generation potential using lifecycle analysis techniques, comes out quite favorably for oceanic mining.[26] However, biodiversity impacts are a source of immense anxiety for environmentalists, and opposition to extraction has gained momentum. There may well be bauxite deposits in the ocean, but the economically viable extraction currently under discussion is for the "battery metals" of nickel, cobalt, manganese, and copper.

The oceans remain a vast but vulnerable frontier for scientific and industrial inquiry. Although often neglected as an immense sink for pollution and plunder, some of the earliest international environmental regulations also involved the oceans. The International Agreement for the Regulation of Whaling was signed in 1937 and is considered the first major multilateral environmental treaty. This agreement was further strengthened by the establishment of the International Whaling Commission and the formalization of the International Convention for the Regulation of Whaling in 1946. Despite the treaty seeing ebbs and flows of national membership over the past several decades, the resilience and recovery of the world's largest mammals has been remarkable. A recent study of recovering whaling populations, published in the *Proceedings of the Royal Society*, showed that the humpback whale population in the southern oceans has gone from a low of four hundred in the 1950s to over 25,000 in 2019.[27]

Nevertheless, the Anthropocene continues to bring new challenges and opportunities for ocean conservation. An immediate threat that has garnered world attention with demonstrable impacts is the scourge of plastics pollution, a mounting menace for oceanic fauna. However, the environmental-activist community is also stridently concerned about another potential longer-term threat—the advent of deep-sea mining. A coalition of over eighty NGOs released a report in May 2020 on what was presented as a review of the scientific literature[28] on the potential harm of seabed mining activity, which is being considered under the UNCLOS. The activists present well-intentioned renderings of "the precautionary principle," which came out of the Earth Summit in 1992 as a means of always erring on the side of caution when potential harms to ecology were unknown.[29] Such an approach in the case of oceanic minerals has received stellar endorsements from doyens of oceanic conservation such as Sir David Attenborough and Sylvia Earle. However, precaution operates on a spectrum, as with any human endeavor, and the principle cannot be used as an excuse for indefinite inertia in a world experiencing competing resource challenges. Caution is in order, but indeterminate precaution is an untenable postulate that can lead to societal paralysis. The case of oceanic minerals has broader applicability in how we might consider decision making to reconcile industrial development with our aspirations for sustainability.

For functional purposes, a systems-science approach is needed to consider the way forward as the deep-sea-mining enterprise has to optimize four overarching and potentially competing objectives:

1. The comparative environmental and social impact of mining on terrestrial versus oceanic ecosystems
2. Supply projections for key critical metals from terrestrial ore reserves (economically viable deposits) and resources (geologically available but currently not economically viable deposits) and from recycled sources and more resource-efficient practices
3. Demand projections for critical metals needed for the transition to cleaner energy sources such as wind and solar power as well as for batteries for electric vehicles and smart grids
4. Revenue-generation potential for sponsoring Small Island Developing States (SIDS) who are partnering with industry for deposit exploration

Much of the debate on oceanic minerals is also couched in terms of their use for green technologies such as wind- and solar-power infrastructure or electric car batteries. The imperative to build such infrastructure comes from concerns about human impacts on the planet's most critical life-sustaining indicators. The time sensitivity of the tipping points in this discourse on planetary boundaries makes the need for technology development more urgent. Hence the option of waiting for alternative technology development and the accumulation of recycled stocks of metals to be available for eventual circularity become less plausible. An increase in recycled stocks is dependent on product durability, whose benefits also need to be measured and compared with the need for more mined inputs through techniques like lifecycle analysis (LCA; discussed in chapter 2). Climate-change mitigation factors into such analyses, as does oceanic ecology, and thus the tradeoffs of sourcing metals to meet demand need to be considered. Within this context, only enough metal so as to have a sustainable stock for recycling should be extracted, ultimately leading to a circular economy in the sector. The modularity of metal uses in electric car batteries makes such circularity plausible. Figure 8.4 presents a heuristic diagram of how we might consider an industrial-ecological approach to decision making at the level of planetary systems.

The case of oceanic minerals provides us an opportunity to consider the complexity that future decisions around resourcing materials will entail as we adapt to a changing climate while trying to advance technologies for human development. What constitutes "development" or "well-being" will be contested at some level, but the global consensus around key parameters like sustainable-development goals or what may follow them in the post-2030 agenda will continue.

THE MATERIALITY OF HEALING

In 1951, President Harry Truman established the Materials Policy Commission (also known as the Paley Commission, after Chairman William Paley) to study the country's natural-resource needs. The mandate of the commission was to look into the long-range outlook for requirements and supplies, helping guard against shortages in the event of another widespread war. The

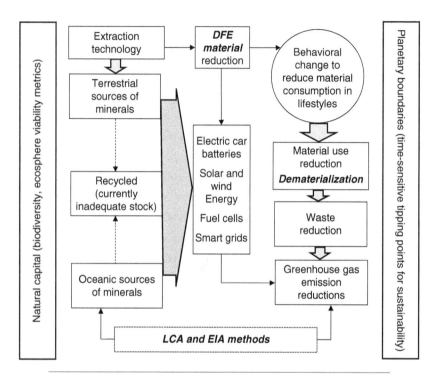

FIGURE 8.4 Applying industrial-ecology approaches to evaluating the viability and need of oceanic mineral extraction. The systems approach considers natural capital and key time-sensitive tipping points as the limiting "walled" parameters for decision making. Industrial ecology methods are italicized.

report it produced, *Resources for Freedom: Foundations for Growth and Security*, paved the way for our broader policy engagement with resource scarcity of materials.

In 1973, the year I was born, the report of the congressionally mandated National Commission on Materials Policy appeared. It was entitled *Materials Needs and the Environment Today and Tomorrow* and contained a detailed discussion of the material cycle and many recommendations for government action. Shortly thereafter, a major Nuclear Regulatory Commission study was published; this monumental effort was a product of the ad hoc Committee on the Study of Materials, chaired by Morris Cohen (COSMAT). The three summary directives for policy makers of these volumes are as follows:

1. Strike a balance between the need to produce goods and the need to protect the environment by modifying the materials system so that all resources, including environmental resources, are all paid for by users.

2. Strive for an equilibrium between the supply of materials and the demand for their use by increasing primary production and by conserving materials through accelerated waste recycling and greater efficiency of use.

3. Manage materials policy more effectively by recognizing the complex interrelationships of the material-energy-environment system, so that laws, executive orders, and administrative practices reinforce policy and not counteract it.

At the time of these efforts, we did not have the full scope of ecological challenges under discussion in policy circles, nor was there as strident an opposition to the extractive enterprises that provided the primary resources. At the turn of the millennium, the mining industry recognized that as the primary resource provider for most materials in durable human use, they needed to have a restorative engagement with broader society. This effort at a "reset" was titled the Mining, Minerals, and Sustainable Development (MMSD) initiative. Although funded by the mining industry, it was a multiactor, multilevel experiment in world politics initiated by a coalition of private-sector firms to respond to this resistance through stakeholder engagement.

Attempts at democratizing global governance are prone to charges of co-optation and were particularly acute in the case of MMSD. Many environmental organizations boycotted the initiative and wrote an open letter to the industry indicating their reasons for that decision. They largely predicated their resistance on the perception that the outcome of the process had been predetermined by the funders, who would characterize mining as "sustainable" under mildly mitigating circumstances. Some of the NGOs that have resisted this effort had an uncompromising normative stance with regard to mining as being inherently unsustainable and thus would label any attempt at defining "sustainable mining" as "greenwashing."[30] There were others, however, who had direct process-oriented concerns about the initiative, and these were able to get some specific workshops organized, such as ones on the rights of indigenous people in mining areas, and they subsequently joined the initiative. Most of the groups that boycotted the main MMSD initiative did, however, attend the culminating conference in Toronto in May 2002, to which I was also invited. While their presentations were not conciliatory by any means, there was at least an engagement of stakeholders during this event. Such engagement may lead to "constructive confrontation" whereby an alternative and more agreeable future from the one individual parties are envisaging is achieved.

Twenty years later, as I reflect on what has changed in charting our material future, the notion of "healing" for the past indiscretions of the extractive industries is what stands out. Healing is linked to what is often termed "restorative justice,"[31] and it is not just for human beings but also connotes a planetary notion of propriety. The term "mineral" is often used to define various naturally occurring materials extracted from Earth's geological crust. Yet, for questions of sustainability, we need to differentiate metallic minerals from nonmetallic ones, because "renewability" for operational purposes will be defined by the level of entropy (or disorder) that mineral use will generate. The question of renewability from a chemical perspective is simply one of expending enough energy to bring back the material from a higher level of entropy to allow for reuse or recycling. Energy needed to counter the entropy created by the mineral's use is the main metric to evaluate whether that material's use is sustainable. From an operational perspective, metallic minerals are used at lower levels of entropy. That is why we are usually able to recycle them, whereas with minerals like coal, the use itself converts the material to such a high level of entropy (in the form of carbon dioxide) that it is essentially nonrenewable.

If we can design products that can retrieve minerals in usable form with a relatively low expenditure of energy and with a restorable environmental impact (particularly if the energy utilized for recycling is from renewable sources), then mineral usage is indeed sustainable. From an economic perspective, the extraction process of a finite resource from Earth's crust can still lead to sustainable development so long as the capital generated is invested in building a diversified economy. This "weak sustainability" aspect also applies to fossil fuel–extraction economies, thus deeming them nonrenewable but sustainable. Indeed, the natural-resource base of some areas may necessitate mineral profits as the catalyst for a longer-term development path. Rather than a simplistic rejection of minerals as "nonrenewable," we must be willing to grapple with the chemical, ecological, and economic nuances of material extraction.

The rise of the aluminum industry gives us an intriguing example of how such nuances can be applied to practically consider the tradeoffs of materiality. As noted in the title of this book, this industry's rise has been unprecedented in its scale and scope. This led to some rash decisions but also created many opportunities for industrial innovation. While there are limits to win-win outcomes for all stakeholders with most industrial development, the aluminum sector has been on a quest for sustainability that is instructive in our broader relationship with natural resources. That quest is ongoing, and by learning more deeply about the natural and social science of our relationship with this remarkable element, we can move more constructively forward in defining our material future.

EPILOGUE

Governing Our Planet's Elemental Resources

An important function of almost every system is to ensure its own perpetuation.

— Donella Meadows, *Thinking in Systems*

I n February 2022, during the final days of completing this manuscript, there was a major convulsion in global commodity chains sparked by the Russian invasion of Ukraine. Physically the world's largest country, Russia is endowed with massive mineral wealth. For much of the twentieth century, these vast resources were governed through the centralized economic system of the Soviet Union. After the fall of the Soviet empire, a mismanaged privatization process took hold. Commodity wealth of the country's vast geological resources was appropriated by those who were willing to take high risks for high rewards. The rise of the infamous Russian "oligarchs" was largely a result of natural capital being captured by these elites, who became power brokers with the state.

The story of the Russian aluminum sector and the state-owned company Rusal is an instructive example of the power of mineral wealth. Between 1991 and 1994, companies set up by a maverick entrepreneur in his twenties, Oleg Deripaska, with the backing of the Uzbek-born Israeli industrialist and entrepreneur Michael Cherney, invested in shares of the Sayanogorsk Aluminum Smelter (SAZ), one of the most technically advanced aluminum plants built by the Soviets. By 2000, Deripaksa's investments managed, among other aluminum-related assets, majority interests in the Sayanogorsk aluminum smelter; the Sayanal foil mill; a fabricating plant in Samara, Russia; and a minority interest in the Nikolaev alumina refinery in Ukraine. There was a relatively lawless period of commodity conflict during the Yeltsin post-Soviet era, during which several oligarchs wrestled for control of the prized mineral assets, in what is referred to as the "aluminum wars."[1] It is widely believed that Deripaska emerged victorious in this war thanks to his marriage to Polina Yumasheva, the daughter of Boris Yeltsin's chief of staff, and his close friendship with Vladimir Putin. Yet another notable oligarch, Roman Abramovich (later to become owner of Chelsea Football Club), sold Deripaska a 50 percent share in Rusal in 2003 and the balance of the company in October 2004. At this point Rusal produced 70 to 80 percent of the Russian aluminum output (2.7 million tons of primary aluminum), which was more than the entire U.S. output of 2.5 million tons and second only to China's production. Globalization allowed for the company to attract foreign investment from companies like the Swiss commodity giant Glencore and the incorporation of Rusal formally as a company in the island tax haven Jersey, a self-governing Crown Dependency of the United Kingdom.

I wondered what would happen to Rusal with the wave of sanctions that the West was imposing and whether this would affect aluminum supply, pricing, and perhaps even demand and use in specific products. After a call with a friend who had worked in the sector for decades, I was told that Rusal had a record of enduring economic shock and would likely adapt. In August 2009, there was a massive accident at Russia's largest hydroelectric dam, the Sayano-Shushenskaya dam in Siberia, which provided 15 percent of Russia's hydroelectric power and 2 percent of its overall power; 70 percent of its output was dedicated to Rusal's use. At least seventy-five people died in the accident, and the repairs cost over $1.5 billion, but despite the short-term challenges, the company bounced back. Some years later, Rusal became the largest corporation to be sanctioned by the U.S. Department of the Treasury, because of the number of shares owned by Deripaska, who had been blacklisted by U.S. lawmakers for his alleged involvement in financing cyberattacks. Such a drastic move sent shockwaves across the industry, and after swift lobbying pressure, the sanctions against Rusal were revoked in 2019 after a reduction in Deripaska's share ownership. The primacy of aluminum supply, as with other commodities, was such that Tesla was still sourcing the metal from Rusal, and no direct sanctions against Rusal have yet been reapplied more than a month into the invasion.[2]

Meanwhile, at an industrial park about an hour's drive from Ho Chi Minh City, a narrow mound of aluminum metal stretching over a kilometer in length was being hoarded by the Vietnam government. Worth an estimated $5 billion, this massive stockpile of metal was equivalent to the entire annual consumption of India, the world's second-most-populous country. The Vietnamese authorities say it was acquired from China by Global Vietnam Aluminum Ltd., known as GVA, and seized as part of a U.S. antidumping investigation against the Chinese billionaire Liu Zhongtian, the chairman of the aluminum giant China Zhongwang Holdings. Previously, this stockpile had been stored in the Mexican desert but was slowly moved to Vietnam. On April 11, 2022, six Southern California companies linked to

Zhongtian were ordered to pay $1.83 billion in restitution for participating in a conspiracy to defraud the United States through a scheme in which huge amounts of aluminum—disguised as "pallets" to avoid $1.8 billion in customs duties—were exported to the United States and "sold" to fraudulently inflate a China-based company's revenues and deceive investors worldwide.[3] The quest for industrial sustainability, with which I had used aluminum as a parable at the start of this book, was being tested in this time of conflict, but such crises made the cause of improved governance even more urgent.

STEWARDSHIP BEYOND BORDERS

Amid such chaos, one may feel disheartened about supply-chain security and mineral governance at multiple levels. To retain some hope, I reached out to the Aluminium Stewardship Initiative (ASI), an organization that had endeavored to bring global aluminum producers, users, and stakeholders all under a banner of improved performance of material supply chains. Aluminum is a metal that has garnered remarkable levels of geopolitical coherence across disparate ideologies, so much so that many facilities in even China and Russia had signed on to improving performance and "chain of custody" through independent assurance. ASI had shepherded this process of voluntary certification, which had largely been prompted by downstream customers asking for greater vigilance of supply chains, from mines to markets, to ensure environmental and social performance. The organization had its origins in a 2009 gathering of stakeholders from the aluminum industry, civil society, academia, and industrial users of aluminum, which had resulted in a Responsible Aluminum Scoping Phase.[4]

The report also underscored the need for an international multistakeholder approach that could complement existing sustainability programs throughout the aluminum industry. This finding ultimately led to the establishment of ASI.

The International Union for the Conservation of Nature (IUCN) was invited by ASI to convene the first standard-setting group in 2012. Within a decade, ASI now has 236 members in six different classes across the supply chain of aluminum, spread across forty-four countries, including China and Russia. They have both a chain-of-custody standard and a performance standard spanning a range of key environmental- and social-performance criteria. Unlike some industry-led certification efforts, which often get dismissed as "greenwashing," ASI has been committed to having strong representation from civil society groups such as the World Wildlife Fund and the Institute for Human Rights and Business.

The Institute for Responsible Mining Assurance (IRMA) has carried the engagement mission a step further by involving nonindustry stakeholders in the governance of the standard. IRMA's decision to certify at varying levels of performance is made by the governance board, which comprises two representatives each from six different stakeholders: (1) mining companies, (2) downstream companies that buy mined products, (3) nongovernmental environmental and social advocacy organizations, (4) organized labor, (5) representatives from impacted communities, and (6) investment and finance houses. Board members aim to make decisions by consensus. However, where this cannot be achieved, there is a vote. Any result with two "no" votes from the same sector does not pass, and the issue must go back to the full group for further discussion and resolution. Thus, a decision cannot pass if one of the stakeholder groups is fundamentally opposed. Aimee Boulanger, who has been the key driving force behind IRMA and now leads the organization, has built an architecture of compliance assurance that has credibility and is also accepted by industry executives. In a heavily male-dominated industry, Aimee has managed to make an impact, which is an encouraging sign of how gender stereotyping around the sector is slowly abating.

Although the Aluminum Stewardship Initiative does not have the same governance structure as IRMA, it has been successful in gaining wide acceptance. Their chief executive officer, Fiona Solomon, credits some of her success to a technocratic culture within the aluminum industry that has enabled her as an academically inclined professional with degrees in engineering and a doctorate in the philosophy of technology to be embraced by the sector. Based in Melbourne, Australia, she endures time-zone challenges and manages an organization with global reach, and she does so with stoic composure and sharp conviction about the impact she can have. Fiona recognizes that change requires planning and persistence. For ASI, she developed a "theory of change" diagram, available on their website—a rare feature with most industry efforts but indicative of her methodical approach to sustainability transitions. Change is inherently incremental, even allowing for variations in pace and magnitude. The external environment is constantly in flux, and the choice of materials for various uses can also change. I asked Fiona about the travails of the sector with reference to energy usage and health impacts, which she acknowledged could mitigate consumer confidence. However, the industry has been nimble in working through such constraints, as evidenced with the massive push for recycling and a circular economy and the targets for reaching carbon neutrality.

The industry has used the term "Practical Minimum Energy" (PME) to predict how new technologies may help it reach ambitious policy targets. The theoretical minimum energy for chemically transforming a material is based on the net chemical reaction used to manufacture the product—in the case of aluminum made from alumina, the theoretical minimum energy is 9.03 kWh per kilogram. The PME lies somewhere between the current best available value and the theoretical minimum value for the reaction. In 2001, the main corporate associations of the sector issued *The Aluminum Industry Vision: Sustainable Solutions for a Dynamic World* and selected a goal of 11 kWh per kilogram for 2020. This represented a 27 percent reduction over the 1995 value of 15.4 kWh per kilogram. As of this publication, I have still not been able to firmly ascertain if the industry has met this target on average; it may well have met it at specific locations or even nationally. Instead of energy efficiency per se, the industry has started to focus on tracking carbon emissions, which has some correlation with efficiency but can also distract from overall energy-consumption metrics in the case of low-carbon sources. Although there is no formal definition of "low-carbon aluminum," ASI has noted that four tons of CO_2 equivalent per ton of aluminum is considered the best achievable with current technologies and accounting for Scope 1,2,3 emissions (direct, energy, and supply-chain emissions, respectively), from "mines to metals," under the Greenhouse Gas Protocol accounting system.[5]

One could also look outside the box at unconventional technologies such as carbothermic reduction (CT) of alumina and chloride reduction of kaolinite clays, which are the only nonelectrochemical processes that have shown potential for aluminum. The theoretical minimum energy for CT comes to around 7.51 kWh/kg. A key benefit of the CT reactor process is the use of electric furnaces, which provide up to 90 percent thermal efficiency; heat loss is limited to conduction and radiation losses from the furnace shell. With these reductions, 37 percent of energy could be saved in comparison with the Hall-Héroult process. Such a process could reduce capital costs by 50 percent or more as a result of volumetric processing through high-intensity smelting. However, upscaling such efforts has still been elusive given high upfront investment costs when established infrastructure exists for conventional processes.[6]

Such technologies are being embraced by a variety of companies, and even those that have not yet joined ASI formally are showing promise of performance improvements. For example, India's Vedanta Aluminum is now included in the Dow Jones Sustainability Index (DJSI) Top 10 for 2021, landing at the fourth spot overall. This index evaluates over ten thousand corporations across several benchmarks measuring economic, environmental, and social criteria. The company was particularly conscientious in 2021 in cutting greenhouse gases by 21 percent to date from fiscal year 2011–2012 levels and recycling 16.5 billion liters of water last year, as well as purchasing 308 million units of renewable power for producing aluminum. So far, 93 percent of the areas mined out by Vedanta have been rehabilitated (partly by using ash generated in operations), and 4 million GJ in energy have been saved by the firm via conservation efforts.[7]

Despite all these investments in greening, the industry is still prospering. Alcoa posted a record high annual net income of US$429 million in 2021 and turned in the highest yearly revenue since 2018 of US$12.2 billion, up by 31 percent on the year. Adjusted earnings rose by 140 percent to US$2.8 billion, and the firm reduced its total debt load to US$1.8 billion in 2021. Meanwhile, even in the fossil-fuel-heavy Gulf region, Emirates Global Aluminum and GE Gas Power joined forces to develop a roadmap to reduce greenhouse-gas emissions by exploring hydrogen as a fuel, as well as carbon capture, utilization, and storage solutions.[8]

Ultimately, the ascendance of one particular material requires multiple tradeoffs. Energy usage and sourcing are key indicators of impact but often not the main determinant in consumer or policy decisions. In a final comment during my interview with her, Fiona Solomon noted how aluminum cans had been falling out of favor at one point, with recyclable plastics gaining traction for beverage containers. However, the rapid concern over the impact and persistence of plastics in the past decade and the global push for a treaty on mitigating plastics pollution in 2022 have made aluminum cans look more favorable again. Ultimately, technical tools such as lifecycle analysis will need to be constantly refined to identify the comparative impacts of materials, and society will need to adapt accordingly. At the same time, governance of supply chains from local communities all the way to international institutions must continue alongside stewardship initiatives.

GLOCALIZATION OF MATERIAL GOVERNANCE?

Much of the history of mineral development has been plagued by conflicts because local communities have disproportionately borne the brunt of impact but not adequately accrued the rewards of development. As I reflected on key lessons in this vein, I was reminded of my conversations in 2014 with the former CEO of one of Australia's largest aluminum companies, Nick Stump, who had made it his mission to change the reward and ownership structure of mineral investment. At the time, Mr. Stump was retired and spent most of the year sailing around the South Pacific on his yacht. However, when an opportunity came up to invest in a new bauxite deposit in northern

Queensland, he came ashore to see if he could make a transformative difference.

Stump signed an indigenous land use agreement with the Wik Way Aboriginal people in relation to the bauxite deposit, to deliver about 15 percent of the profits to the local community, the council, and the traditional owners. This was his effort at addressing the wrongs of the past. As he noted in one interview: "The point worth making is that after 60 years of mining [on the Cape York peninsula] these people are still living in squalor." All they had received were small gestures of engagement, such as the Kurri Kurri smelter and beef-farm buffer zone of 16,000 hectares of bushland to absorb escaped fluoride emissions. Beehives were kept to monitor fluoride levels.

Aboriginal people had been engaged in a struggle for their rights in bauxite-mining regions for decades. In 1963, the first direct assertion of indigenous claims to land in Australia was linked to bauxite development in the Gove Peninsula on the land of the Yolngu people. The petitions from the community of Yirrkala were beautifully framed in decorated tree bark and presented to parliament at the time, but with minimal impact. It was not until 1978 that the Yolngu claims were formally recognized under the Aboriginal Land Rights Act (Northern Territory), which had been passed two years earlier.

By giving the people a substantive share in the project and with a far more generous community-development package than was the norm, Stump wanted to leave a legacy of positive change. However, when the final decision was made by the state government of Queensland, the project was offered instead to the Swiss conglomerate Glencore rather than the Aboriginal Aurukun Bauxite Corporation. As one community member later stated, the government came in "like the KGB, the FBI and the CIA all rolled into one" and thwarted the bid. The Wik people of Cape York sued to overturn laws that allowed the government to supersede their preference, but they lost in high court. As I was finishing this manuscript, I did a quick news search to see if there had been any development since 2015 on the Aboriginal efforts to regain control. The bauxite project in question is now listed on Glencore's site after being granted a mineral development license in 2018 and is under development. There has since been a glimmer of hope, with the opening of an indigenous-owned bauxite mine in Arnhem Land, which is owned by the Yolngu Aboriginal community; it produced its first ore for export to China in 2017. Two years later, it was also granted ASI certification. This may indeed lead to the kind of transformative change that Stump envisaged for the community.

The extractive sector has the potential to function as a catalyst for development in mineral-rich countries, but many challenges prevent this potential from being fully realized. One crucial hurdle is to link local governance and control efforts, such as what Stump and the Aurukun community were envisaging, with global norms of mineral governance. Relevant challenges to governance include illicit financial flows, lack of transparency and accountability and the associated risk of corruption, political instability, global asymmetries of power and conflicting stakeholder interests leading to social conflict, and conflict between formal and informal mining activities. Key governance gaps include unintended consequences of governance instruments that undermine the sustainability of mining, lack of buy-in to existing instruments, lack of compliance with existing instruments, uneven focus of current instruments relative to the broad range of issues confronting the extractive sector as a whole, proliferation of standards concerning various aspects of mining sustainability, and the lack of a coherent and collective theory of change.

Perhaps a hybrid vision combining local and global norms could be instructive in planning for more effective governance of materials. The term "glocalization" has its origins in the Japanese concept of *dochakuka*, which means "global localization." Such an approach was the one adopted by Japanese industry's management in an attempt to balance local cultural proclivities with a globalized customer base. Formally, glocalization is defined by Joachim Blatter in the *Encyclopedia of Governance* as "simultaneous occurrence of both universalizing and particularizing tendencies in contemporary social, political, and economic systems."[9] We need the development of a new governance framework that involves all actors in the extractive sector and whose implementation is a responsibility shared by host and home countries along the extractive-value chain. The framework should integrate all pillars of sustainable development—people, planet, prosperity, peace, and partnership—and set out principles, policy options, and good practices for enhancing the extractive sector's contribution to achieving the UN Sustainable Development Goals by 2030.

The value of a glocalization approach is particularly salient with reference to critical materials sourcing for a green transition. Fast-emerging technologies that use minerals, such as electric-car batteries, solar panels, wind turbines, communication devices, and military applications, are expected to drive tremendous growth and demand for these metals in the near future. Yet we have no global mechanism for governing material flows. Circular-economy principles such as eco-design; recycling, refurbishment, and reuse; and development of secondary sources of minerals and metals are considered promising options for the near-future supply of what is sometimes termed Technology Critical Elements (TCEs). The amount of TCEs recycled from e-waste is not yet significant, attributable to low critical element concentrations in waste flows, dissipative applications, and the critical element being a minor composition in a complex material matrix, among other factors. The widespread application of TCEs in different industries and in agriculture is expected to increase the concentrations of these elements in the environment, which would disturb not only aquatic systems but also plant and soil ecosystems, leading to a range of human health issues. Close monitoring, therefore, is needed at places where phosphate-based fertilizers are used, in areas where soil conditions are favourable to TCE mobility, where there is availability and uptake by plants, and at e-waste dump sites where surface runoff could contaminate the local environment. Extensive use of TCEs in day-to-day life requires the development of human, as well as technical, capacity to undertake toxicological assessment of these elements from a human health perspective. As noted in a recent report by the International Energy Agency, the green energy transition necessitates a "shift from a fuel-intensive to a material-intensive energy system." Resourcing this transition will require an increase in minerals production of lithium, graphite, nickel, and cobalt rare earths by 4,200 percent, 2,500 percent, 1,900 percent, and 700 percent, respectively, by 2040.[10] How such material intensity might be managed will be one of the defining challenges of our times in years to come.

THE FACTOR "X" CODA

In the year of my birth, 1973, a National Commission on Materials Policy published a series of reports prompted by concerns similar to what we are experiencing today with reference to materials criticality. James Boyd, the executive director of the commission gave a lecture on the topic titled *The Resource*

Trichotomy, referring to materials, energy, and the environment. This was subsequently revisited by the former director of the National Institutes of Standards and Technology Lyle Schwartz in a paper in 1998 in which he presented a diagram on materials and sustainability (figure E.1).

The "materials cycle" can only keep up with rising demand if we can find ways of being more efficient at resource usage without succumbing to a rebound effect of increased demand. The same year that Schwartz published his paper, the Organization for Economic Cooperation and Development (OECD), which now boasts a major annual Responsible Minerals Forum, adopted a long-range goal to decrease materials intensities by a factor of ten over the next four decades. This would be equivalent to using only 66 pounds of materials per $100 gross domestic product, compared to the present value of approximately 660 pounds per

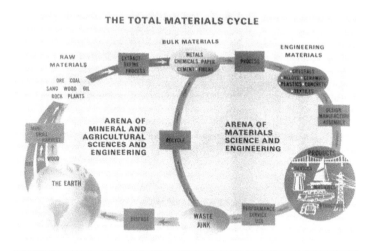

FIGURE E.1 Sustainability: the materials role.

Source: Adapted from Lyle H. Schwartz, "1998 Distinguished Lecture in Materials and Society of ASM International," *Metallurgical and Materials Transactions A* 30 (April 1999).

$100 GDP. Underlying this target setting was what has been often called a "Factor X" approach to materials policy.

The concept has its origins with the German government's planning processes, inspired by the physicist Friedrich Schmidt-Bleek in the 1990s. Bleek, who was affectionately called "Bio," noted that if sustainable development is to be achieved globally and for all of humankind, industrialized countries must reduce their consumption of resources by a factor of 10 (the roman numeral X), or 90 percent, within fifty years. Subsequently he established the World Resources Forum, which began to hold annual summer gatherings in Davos, Switzerland, as a more science-focused shadow event to the World Economic Forum, held at the same venue in winter. In 1995, the German legislator Ernst-Ulrich von Weizsäcker stated that a factor of four can double wealth and halve resource consumption. The efforts of von Weizsäcker were also supported by the Club of Rome—an institution that had emerged from the Limits to Growth modeling enterprise of the early 1970s.[11]

The German Environment Agency still champions the Factor X target-setting rubric on its websites and states that

> increasingly intensive use of natural resources by humans is causing ecosystems to exceed their stress limits and exacerbating global environmental problems. In the last 40 years, the global extraction of raw materials has more than tripled to around 85 billion tons per year. Today, this already significantly exceeds the earth's regenerative capacity and endangers the development opportunities of future generations.

In essence, the concept suggests that the "use of natural resources must become x times more intelligent and efficient." X times more use could be made and therefore x times more wealth be generated from a ton of raw material.[12] Thus a factor of 4 (= 75 percent increase), 10 (90 percent), or more suggests the efficiency increase that could be attained, and the symbolism of "X" implies that this factor has unforeseen potential.

On the social and economic side, the Factor X approach finds an interesting manifestation in what the economist Miles Kimball calls the "Aluminum Rule." Formulated as a less demanding version of the biblical Golden Rule, he states it as follows: "When acting collectively—or considering collective actions—put a weight on the welfare of human beings outside the in-group at least one-hundredth as much as the welfare of those in the in-group."[13] Hence social investments in resource development could be considered in terms of a "Factor 100." I contacted Kimball and asked him why he used the metaphor of aluminum, and he noted that it was simply because in contrast with a rare luxury metal like gold, he wanted to suggest a more practical everyday metal!

By this "rule of thumb," a dollar is one hundred times as valuable to someone who has only one tenth the income. The Aluminum Rule thus alerts us to always keeping an eye on the return on investment in well-being and livelihood when deciding on resource investments. Ultimately, it is such careful targeting and calibration of our resource usage that will keep industry and citizens in general on track toward a more sustainable future.

The elemental aspect of the story of human development is defined by finding effective and efficient ways of converting the material and energy resources of the planet into functional forms. The arrival of aluminum in our lives is a parable for this grander universal epic. Here was a metal that we had not even isolated in pure form until the late nineteenth century—in comparison with iron or copper, which had been known for millennia. Yet the ability of human creativity to connect energy and material demand allowed for its proliferation into myriad products—for better and for worse. We began our journey in this book by understanding how the laws of nature convert energy into matter and then further refine that matter into myriad minerals. Humans in turn convert those minerals into multitudinous materials. Industry is at the heart of this series of great transformations.

As a species, we have defined ourselves by the ability to form industry—the word itself is synonymous with human "work." Other organisms work as well, but human work is different in its scale and ambition. Ants, termites, and coral work and build great architectural marvels, but their work is not defined by a desire to develop markets for products. They do not have to contend with situations where consumerism and morality might clash. Aluminum's story shows us how humans can try to navigate such clashes. Our path may meander from mistakes to monumental achievements, depending on how well we understand the scientific and societal limits of endurance.

NOTES

PREFACE

1. U.S. Geological Service, "Mineral Commodity Summaries 2022—Aluminum," https://pubs.usgs.gov/periodicals/mcs2022/mcs 2022-aluminum.pdf.

2. International Aluminum Institute, *Opportunities for Aluminum in a Post-COVID Economy* (London: IAI [CRU Group], 2022).

1. ELEMENTAL ORIGINS AND THE INVENTION OF NEED

1. J. Kovac and M. Weisberg, eds., *Roald Hoffmann on the Philosophy, Art, and Science of Chemistry*, illustrated ed. (Oxford: Oxford University Press, 2012).

2. K. Fukui, "The Uniqueness of Nature and Human Beings," *International Journal of Quantum Chemistry* 53 (1995): 119–24.

3. S. Weinberg, *The First Three Minutes: A Modern View of the Origin of the Universe* (New York: Basic, 2008).

4. R. Hazen, "Evolution of Minerals," *Scientific American*, March 2010.

5. A. R. Butler, C. Glidewell, and S. E. Pritchard, "Aluminum Objects from a Jin Dynasty Tomb—Can They Be Authentic?," *Interdisciplinary Science Reviews* 11, no. 1 (1986): 88–94.

6. C. Singer, *The Earliest Chemical Industry: An Essay in the Historical Relations of Economics & Technology Illustrated from the Alum Trade* (Folio Society, 1948).

7. A. Günster and S. Martin, "A Holy Alliance: Collusion in the Renaissance Europe Alum Market," *Review of Industrial Organization* 47, no. 1 (2015): 1–23.

8. R. B. Ekelund, R. F. Hébert, and R. D. Tollison, "An Economic Model of the Medieval Church: Usury as a Form of Rent Seeking," *Journal of Law, Economics, & Organization* 5, no. 2 (1989): 307–31.

9. V. Postrel, "Not Even the Pope Can Maintain a Monopoly," *Reason*, April 2019.

10. UK National Trust for Yorkshire, https://www.nationaltrust.org.uk /yorkshire-coast/features/how-alum-shaped-the-yorkshire-coast.

11. S. Chance, *The Alum Maker's Secret* (Wiskard Books).

2. SOIL WITHOUT SOUL: WHY ALUMINUM WAS REJECTED BY LIFE

1. S. Duval et al., "Minerals and the Emergence of Life," in *Metals, Microbes, and Minerals: The Biogeochemical Side of Life*, ed. Kroneck and Torres (De Gruyter, 2021), 135–58.

2. M. J. Russell, W. Nitschke, and E. Branscomb, "The Inevitable Journey to Being," *Philosophical Transactions of the Royal Society B: Biological Sciences* 368, no. 1622 (2013).

3. A. M. Soto, C. Schaeberle, M. S. Maier, et al., "Evidence of Absence: Estrogenicity Assessment of a New Food-Contact Coating and the Bisphenol Used in Its Synthesis," *Environmental Science & Technology* 51, no. 3 (2017): 1718–26.

4. Food Packaging Forum, https://www.foodpackagingforum.org/food -packaging-health/can-coatings.

5. "Packaging for Good, Inside and Out," *National Geographic*, https:// www.nationalgeographic.com/environment/article/partner-content -packaging-for-good.

6. C. Exley, *Imagine You Are an Aluminum Atom: Discussions with Mr. Aluminum* (Skyhorse, 2020).

7. I. Krupińska, "Aluminum Drinking Water Treatment Residuals and Their Toxic Impact on Human Health," *Molecules* 25, no. 3 (2020): 641.

8. E. Fermi, "Artificial Radioactivity Produced by Neutron Bombardment," Nobel Foundation, 1938, https://www.nobelprize.org/uploads /2018/06/fermi-lecture.pdf.

9. N. D. Priest, "The Biological Behaviour and Bioavailability of Aluminum in Man, with Special Reference to Studies Employing Aluminum-26 as a Tracer: Review and Study Update," *Journal of Environmental Monitoring: JEM* 6, no. 5 (2004): 375–403.

10. D. F. Ciambrone, *Environmental Life Cycle Analysis*, 1st ed. (CRC, 2018).

11. R. J. Mitkus, D. B. King, M. A. Hess, et al., "Updated Aluminum Pharmacokinetics Following Infant Exposures Through Diet and Vaccination," *Vaccine* 29, no. 51 (2011): 9538–43.

12. J. D. Birchall, C. Exley, J. S. Chappell, and M. J. Phillips, "Acute Toxicity of Aluminum to Fish Eliminated in Silicon-Rich Acid Waters," *Nature* 338, no. 6211 (1989): 146–48.

13. A. Zioła-Frankowska, Ł. Kubaszewski, M. Dąbrowski, and M. Frankowski, "Interrelationship Between Silicon, Aluminum, and Elements Associated with Tissue Metabolism and Degenerative Processes in Degenerated Human Intervertebral Disc Tissue," *Environmental Science and Pollution Research International* 24, no. 24 (2017): 19777–84.

14. H. G. Wells, "Another Basis for Life," *Saturday Review*, December 22, 1894.

15. J. E. Reynolds, in *Nature* 48, no. 477 (1893). He built his argument on earlier work, such as R. S. W. Ball, "The Possibility of Life in Other Worlds," *Fortnightly Review* 62, no. 56 (1894); and A. Alison, "Possible Forms of Life," *Journal of the British Interplanetary Society* 21, no. 48 (1968).

16. D. Darling and D. Schulze-Makuch, *The Extraterrestrial Encyclopedia*, 1st ed. (Design Publishing, 2016).

3. UNBREAKABLE BONDS:
THE CHALLENGE OF EXTRACTION

1. E. Kinne-Saffran and R. K. Kinne, "Vitalism and Synthesis of Urea: From Friedrich Wöhler to Hans A. Krebs," *American Journal of Nephrology* 19, no. 2 (1999): 290–94. https://doi.org/10.1159/000013463.

2. P. J. Ramberg, "The Death of Vitalism and the Birth of Organic Chemistry: Wöhler's Urea Synthesis and the Disciplinary Identity of Organic Chemistry," *Ambix* 47, no. 3 (2000): 170–95. https://doi.org/10.1179/amb.2000.47.3.170.

3. J. H. White, *The History of the Phlogiston Theory* (London: Edward Arnold, 1932).

4. "White Aluminum Pavilion Is Inspired by Author Jules Verne," MaterialDistrict, https://materialdistrict.com/article/white-aluminum -pavilion-jules-verne/.

4. THE BOND BREAKERS AND THEIR BOUNTY

1. "Aluminum and Its Manufacture by the Deville-Castner Process," *Science* 13, no. 322 (1889): 260–65; W. Anderson, "Aluminum and Its Manufacture by the Deville Process. Report of the Fourteenth Ordinary Meeting, March 13, 1889," *Journal of the Society of Arts*, March 15, 1889.

2. J. Lockwood, "Topping of the Tip: How Aluminum Found Its Way Onto the Washington Monument," *Prologue*, Summer 2014, https://www.archives.gov/files/publications/prologue/2014/summer/aluminum.pdf.

3. G. J. Binczewski, "A History of the Aluminum Cap of the Washington Monument," *Journal of Minerals* 47, no. 11 (1995), https://www.tms .org/pubs/journals/jom/9511/binczewski-9511.html.

4. M. Laparra, "The Aluminum False Twins. Charles Martin Hall and Paul Héroult's First Experiments and Technological Options," *Cahiers d'histoire de l'aluminum* 48, no. 1 (2012): 84–105.

5. John A. S. Green, "Aluminum Recycling and Processing for Energy Conservation and Sustainability," Materials Park Ohio ASM International, 2007.

6. Quoted in J. Edwards, *The Immortal Woodshed: The Story of the Inventor Who Brought Aluminum to America* (New York: Dodd, Mead, 1955).

7. M. M. Trescott, ed., *Dynamos and Virgins Revisited: Women and Technological Change in History: An Anthology* (Metuchen, NJ: Scarecrow, 1979).

8. Letter from Emily Acton Philipps to *Invention and Technology* 14, no. 3 (1999).

9. C. Flavell-White, "Paul Heroult and Charles Hall: Turning a Rarity into a Commodity," *The Chemical Engineer*, June 1, 2013, https://www .thechemicalengineer.com/features/cewctw-paul-h%C3%A9roult -and-charles-hall-turning-a-rarity-into-a-commodity/.

10. I. Peritz, "Town That Aluminum Built Hopes to Join World Heritage List," *Globe and Mail*, November 12, 2010, https://www.theglobeandmail.com/news/national/town-that-aluminum-built-hopes-to-join-world-heritage-list/article1461469/.

11. City of Saguenay, Universal Heritage Value Statement for Arvida, https://arvida.saguenay.ca/en/reconnaissance-patrimoniale/heritage-br-recognition-process/the-universal-heritage-value-of-arvida.

12. J. M. Hartwick, *Out of Arvida* (Citoxique, 2007).

13. Letter to the editor, *Globe and Mail* (Toronto), November 15, 2010, https://www.theglobeandmail.com/opinion/letters/nov-15-letters-to-the-editor/article1314231/.

14. P. Reed, "Galligu: An Environmental Legacy of the Leblanc Alkali Industry, 1814–1920," RSC ECG, https://www.envchemgroup.com/galligu-an-environmental-legacy-of-the-leblanc-alkali-industry-1814-1920.html.

15. Newcastle upon Tyne Town Council, *Newcastle Council Reports*, 1840, https://books.google.com/books?id=4hZMAQAAMAAJ.

16. K. Lockhart, "How This Abandoned Mining Town in Greenland Helped Win World War II," *Smithsonian*, 2019. https://www.smithsonianmag.com/travel/how-abandoned-mining-town-greenland-helped-win-world-war-ii-180973835/.

17. "Eclipse Discovers Potential of Greenland," *Mining Journal*, July 13, 2021, https://www.mining-journal.com/resourcestocks-company-profiles/resourcestocks/1413656/eclipse-discovers-potential-of-greenland.

5. MOBILE METAL: HOW ALUMINUM FACILITATED WAR AND PEACE

1. "Golden Jubilee of Aluminum," *Science* 83, no. 2151 (March 20, 1936).

2. Quentin R. Skrabec, *Aluminum in America: A History* (McFarland, 2017).

3. George David Smith, *From Monopoly to Competition: The Transformations of Alcoa, 1888–1986*, 1st ed. (Cambridge: Cambridge University Press, 1988).

4. Mimi Sheller, *Aluminum Dreams: The Making of Light Modernity*, ill. ed. (Cambridge, MA: MIT Press, 2014).

5. D. H. Anderson, *Aluminum for Defense and Prosperity* (Washington, DC: Public Affairs Institute, 1951), 3.

6. United States Office of Economic Warfare, *Reoccupation Guide to the Mineral Resources of Greece*, July 1943.

7. NASA, "Aluminum – Rocketology: NASA's Space Launch System," https://blogs.nasa.gov/Rocketology/tag/aluminum/.

8. Darryl Day Spencer, "The Importance of Aluminum Oxide Aerosols to Stratospheric Ozone Depletion," Massachusetts Institute of Technology, 1996, https://dspace.mit.edu/handle/1721.1/41371.

9. Harry Jones, "The Recent Large Reduction in Space Launch Cost," July 2018, https://ttu-ir.tdl.org/handle/2346/74082.

10. Excerpts from the Public Papers of President Dwight D. Eisenhower, speech broadcast on television from the White House on January 17, 1971.

11. *United States v. Aluminum Co. of America*, 2 Cir., 1945, 148 F.2d 416, 424.

12. Robert H. Jackson Center Archives, https://www.roberthjackson.org/lesson-plan/the-andrew-mellon-case-1934-1937/. The center is named for the solicitor general of the United States.

13. "Teddy Roosevelt's Prescient Denunciation of Corporate Power," *Provincetown Independent*, August 18, 2021, https://provincetownindependent.org/history/2021/08/18/teddy-roosevelts-prescient-denunciation-of-corporate-power/.

14. Quoted in Smith, *From Monopoly to Competition*, 196.

15. Ayn Rand, *Capitalism: The Unknown Ideal, with Additional Articles by Nathaniel Branden, Alan Greenspan, and Robert Hessen*, 1st ed. (New York, New American Library, 1966).

16. Ronald Humphrey, *Alcoa and Paul O'Neill's Life-Saving Habits* (London, 2013). https://doi.org/10.4135/9781506324272.

17. R. A. Williams, "The Medieval and Renaissance Origins of the Status of the American-Indian in Western Legal Thought," *Southern California Law Review* 57 (1983): 1.

18. "Guyana Has US$1B in Untapped Bauxite Potential," *Guyana Chronicle*, March 7, 2022, https://guyanachronicle.com/2022/03/07/guyana-has-us1b-in-untapped-bauxite-potential/.

19. Penda Diallo, *Regime Stability, State Insecurity, and Bauxite Mining in Guinea: Developments Since the Late Twentieth Century* (London: Routledge, 2020).
20. "Guinea Achieves High Score in EITI Implementation," Extractive Industries Transparency Initiative, February 16, 2022, https://eiti.org/news/guinea-achieves-high-score-in-eiti-implementation.
21. Stephan F. Miescher, *A Dam for Africa: Akosombo Stories from Ghana* (Bloomington: Indiana University Press, 2022).

6. ALUMINUM FOR ALL: THE INVENTION OF A HOUSEHOLD METAL

1. "Transparent Aluminum Is 'New State of Matter,'" *ScienceDaily*, July 9, 2009, https://www.sciencedaily.com/releases/2009/07/090727130814.htm.
2. *Aluminum World* 6 (February 1900).
3. E. Dash, "David Reynolds, Leader of Metals Company, Dies at 96," *New York Times*, September 1, 2011.
4. Kaiser aluminum and asbestos: https://www.asbestos.com/companies/kaiser-aluminum/.
5. J. Helfand, *Design: The Invention of Desire* (New Haven, CT: Yale University Press, 2016).
6. M. Sheller, *Aluminum Dreams: The Making of Light Modernity* (Cambridge, MA: MIT Press, 2014), 173.
7. Details of the Ford case from the Yale Center for Business and the Environment: https://cbey.yale.edu/research/alcoa-the-auto-industry.
8. "Using Aluminum and Water to Make Clean Hydrogen Fuel—When and Where It's Needed," *MIT Energy Initiative*, https://energy.mit.edu/news/using-aluminum-and-water-to-make-clean-hydrogen-fuel-when-and-where-its-needed/. L. Meroueh et al., "Leveraging Grain Size Effects on Hydrogen Generated Via Doped Aluminum-Water Reactions Enabled by a Liquid Metal," *ACS Applied Energy Materials* 4, no. 1 (January 25, 2021): 275–85, https://doi.org/10.1021/acsaem.0c02175.
9. J. Fikar, "Al-Cu-Fe Quasicrystalline Coatings and Composites Studied by Mechanical Spectroscopy," EPFL, 2003, https://doi.org/10.5075/epfl-thesis-2707.

10. I. Han, K. L. Wang, A. T. Cadotte, et al., "Formation of a Single Quasicrystal Upon Collision of Multiple Grains," *Nature Communications* 12, no. 1 (2021): 5790, https://doi.org/10.1038/s41467-021-26070-9.

11. G. Djukanovic, "Are Aluminum-Scandium Alloys the Future?," *Aluminum Insider*, July 28, 2017, https://aluminuminsider.com/aluminum -scandium-alloys-future/.

12. A. V. Boyarintsev, H. Y. Aung, S. I. Stepanov, et al., "Evaluation of Main Factors for Improvement of the Scandium Leaching Process from Russian Bauxite Residue (Red Mud) in Carbonate Media," *ACS Omega* 7, no. 1 (2021): 259–73, https://doi.org/10.1021/acsomega.1c04580.

13. K. Hund et al., *Minerals for Climate Action* (Washington, DC: World Bank, 2020).

14. N. P. Hylton, X. F. Li, V. Giannini, et al., "Loss Mitigation in Plasmonic Solar Cells: Aluminum Nanoparticles for Broadband Photocurrent Enhancements in GaAs Photodiodes," *Scientific Reports* 3, no. 1 (2013): 2874, https://doi.org/10.1038/srep02874.

15. Y. Kong, C. Tang, X. Huang, et al., "Thermal Reductive Perforation of Graphene Cathode for High-Performance Aluminum-Ion Batteries," *Advanced Functional Materials* 31, no. 17 (2021): 2010569, https:// doi.org/10.1002/adfm.202010569. Summary in "EV Range Breakthrough with New Aluminum-Ion Battery," *Light Metal Age Magazine*, June 17, 2021, https://www.lightmetalage.com/news/industry-news /applications-design/ev-range-breakthrough-with-new-aluminum -ion-battery/.

16. S. Yang and H. Knickle, "Design and Analysis of Aluminum/Air Battery System for Electric Vehicles," *Journal of Power Sources* 112, no. 1 (October 24, 2002): 162–73, https://doi.org/10.1016/S0378 -7753(02)00370-1.

17. R. T. Prider, "Blowpipe Analysis of Minerals," in *Mineralogy* (Boston: Springer, 1983), 47–55, https://doi.org/10.1007/0-387-30720-6_19.

18. K. L. Watkins and L. L. Southern, "Effect of Dietary Sodium Zeolite A on Zinc Utilization by Chicks," *Poultry Science* 72, no. 2 (February 1993): 296–305.

19. A. Mastinu et al., "Zeolite Clinoptilolite: Therapeutic Virtues of an Ancient Mineral," *Molecules* 24, no. 8 (January 2019): 1517, https://doi .org/10.3390/molecules24081517.

20. H. Derakhshankhah et al., "Biomedical Applications of Zeolitic Nanoparticles, with an Emphasis on Medical Interventions," *International Journal of Nanomedicine* 15 (January 21, 2020): 363–86, https://doi .org/10.2147/IJN.S234573.

7. RECYCLING AND REALISM: THE INDUSTRIAL ECOLOGY PARADIGM

1. *Calexico Chronicle*, March 8, 1939, California Digital Newspaper Collection, https://cdnc.ucr.edu/?a=d&d=CC19390308.2.57.
2. *Colorado Transcript*, January 15, 1959, Colorado Historic Newspapers Collection, https://www.coloradohistoricnewspapers.org/?a=d&d =CTR19590115-01.2.8.
3. This data and the following descriptive detail of the innovations to the recycling process developed by Dewan and his associates are documented from internal BRAMCO documents that have been verified by the Bahraini government.
4. E. L. Scheller, B. L. Ehlmann, R. Hu, et al., "Long-Term Drying of Mars by Sequestration of Ocean-Scale Volumes of Water in the Crust," *Science* 372, no. 6537 (March 16, 2021), https://doi.org/10.1126/science .abc7717.
5. P. M. Celliers et al., "Insulator-Metal Transition in Dense Fluid Deuterium," *Science* 361, no. 6403 (2018): 677–82.
6. For a range of resources in this regard, see https://seea.un.org/content /material-flow-accounts.
7. There are now open-source LCA databases as well as commercially available ones. See https://www.openlca.org/lca-data/.
8. U. Mama, "Cana, Benin, an Important piece of Black History," *African Magazine*, January 12, 2017, https://myafricanmagazine.com/cana -a-neglected-piece-of-black-history/.
9. J. Harris, "Planned Obsolescence: The Outrage of Our Electronic Waste Mountain," *Guardian*, April 15, 2020, https://www.theguardian .com/technology/2020/apr/15/the-right-to-repair-planned-obsole scence-electronic-waste-mountain.
10. D. O'Connell et al., "Resilience, Adaptation Pathways and Transformation Approach: A Guide for Designing, Implementing and

Assessing Interventions for Sustainable Futures (Version 2)," Canberra: Commonwealth Scientific and Industrial Research Organization, 2019, https://research.csiro.au/eap/rapta.

11. For an excellent primer on the topic, refer to C. Siskin, *System: The Shaping of Modern Knowledge* (Cambridge, MA: MIT Press, 2016); D. H. Meadows, *Thinking in Systems* (Chelsea Green, 2008).

12. "Largest Electronic Material Recycler in the World—Boliden," https://www.boliden.com/sustainability/case-studies/largest-electronic-material-recycler-in-the-world.

13. U.S. Department of Energy, "U.S. Energy Requirements for Energy Production," 2007, https://www1.eere.energy.gov/manufacturing/resources/aluminum/pdfs/al_theoretical.pdf.

14. É. Lèbre, J. R. Owen, G. D. Corder, et al., "Source Risks as Constraints to Future Metal Supply," *Environmental Science & Technology* 53, no. 18 (2019): 10571–79, https://doi.org/10.1021/acs.est.9b02808.

15. Quoted in M. Sheller, *Aluminum Dreams: The Making of Light Modernity* (Cambridge, MA: MIT Press, 2014).

16. See the dynamic data diagrams on the aluminum cycle at https://alucycle.international-aluminum.org/public-access/#regional.

17. UNEP, "Recycling Rates of Metals—A Status Report. A Report of the Working Group on Global Metal Flows to the International Resource Panel," 2011, https://www.resourcepanel.org/reports/recycling-rates-metals.

8. RESTORATION AND RENEWAL OF MINERAL FRONTIERS

1. "Build Back Better: In the Winter of the 5th Kondratieff Wave," IPR blog, June 11, 2020, https://blogs.bath.ac.uk/iprblog/2020/06/11/build-back-better-in-the-winter-of-the-5th-kondratieff-wave/.

2. "Noranda Opens New Reservoir to Aid Fight Against Bauxite Dust in Jamaica," *Aluminum Insider,* January 9, 2022, https://aluminuminsider.com/noranda-opens-new-reservoir-to-aid-fight-against-bauxite-dust-in-jamaica/.

3. The NASA image visualization can be found at https://svs.gsfc.nasa.gov/2640.

4. M. R. O'Connor, "One of the Most Repeated Facts About Haiti Is a Lie," *Vice* (blog), October 13, 2016, https://www.vice.com/en/article/43qy9n/one-of-the-most-repeated-facts-about-deforestation-in-haiti-is-a-lie.

5. S. Blair Hedges, Warren B. Cohen, Joel Timyan, and Zhiqiang Yang, "Haiti's Biodiversity Threatened by Nearly Complete Loss of Primary Forest," *Proceedings of the National Academy of Sciences* 115, no. 46 (2018): 11850–55, https://doi.org/10.1073/pnas.1809753115.

6. O'Connor, "One of the Most Repeated Facts About Haiti Is a Lie."

7. F. Doura, *Économie d'Haïti. Dépendance, crises, et développement* (Éditions DAMI, 2012). Quoted in Sheller, 2014. M. Sheller, *Aluminum Dreams: The Making of Light Modernity* (Cambridge, MA: MIT Press, 2014).

8. "Haiti and Dominican Republic to Jointly Counter Environmental Degradation and Boost Food Security in Border Zone," UNEP, August 7, 2017, http://www.unep.org/news-and-stories/story/haiti-and-dominican-republic-jointly-counter-environmental-degradation-and.

9. G. Power et al., "Bauxite Residue Issues," *Hydrometallurgy* 108 (2011): 1–2.

10. M. Bagani, E. Balomenos, and D. Panias, "Nepheline Syenite as an Alternative Source for Aluminum Production," *Minerals* 11 (2021): 734, https://doi.org/10.3390/min11070734.

11. Á. D. Anton, O. Klebercz, Á. Magyar, et al., "Geochemical Recovery of the Torna–Marcal River System After the Ajka Red Mud Spill, Hungary," *Environmental Science: Processes & Impacts* 16, no. 12 (2014), 2677–85, https://doi.org/10.1039/C4EM00452C.

12. A. Gelencsér, N. Kováts, B. Turóczi, et al., "The Red Mud Accident in Ajka (Hungary): Characterization and Potential Health Effects of Fugitive Dust," *Environmental Science & Technology* 45, no. 4 (2011): 1608–15, https://doi.org/10.1021/es104005r.

13. University of Helsinki and University of Leuven Red Mud project, https://etn.redmud.org/project/.

14. "Red Mud Becomes a Green Raw Material," KU Leuven Stories, January 21, 2021, https://stories.kuleuven.be/en/stories/red-mud-becomes-a-green-raw-material.

15. University of Geneva, "Novel Solution to Drastically Reduce World's Largest Waste Stream," https://phys.org/news/2022-04-solution-drastically -world-largest-stream.html.

16. "Sustainable Bauxite Residue Management Guidance," International Aluminum Institute, 2022, https://international-aluminum.org /resource/sustainable-bauxite-mining-guidelines-second-edition -2022-2/.

17. K. McRae, "Effects of PCB Contamination on the Environment and the Cultural Integrity of the St. Regis Mohawk Tribe in the Mohawk Nation of Akwesasne," PhD diss., University of Vermont, January 1, 2015, https://scholarworks.uvm.edu/graddis/522.

18. *United States of America, State of New York, and St. Regis Mohawk Tribe v. Alcoa Inc. and Reynolds Metals Co.*, July 17, 2013.

19. Brenda LaFrance, personal communication via email, November 10, 2021.

20. Mohawk Thanksgiving Prayer, http://aihc1998.tripod.com/mohawkpr .html.

21. R. L. Chaney et al., "Phytoremediation of Soil Metals," *Current Opinion in Biotechnology* 8, no. 3 (1997): 279–84.

22. Quoted in *Tehelka*, March 13, 2002, http://tehelka.com/.

23. P. Kumar, G. K. Nigam, M. K. Sinha, and A. Singh, *Water Resources Management and Sustainability* (Springer, 2022).

24. Anil Tripathy et al., "AA23—Preparation of Metallurgical Grade Alumina from Coal Fly Ash," 2020, 4.

25. For a recent summary of the environmental concerns related to deep sea mining, refer to the High Level Panel for a Sustainable Ocean Economy's Blue Paper, June 2020, https://www.wri.org/initiatives /high-level-panel-sustainable-ocean-economy.

26. D. Paulikas et al., "Life Cycle Climate Change Impacts of Producing Battery Metals from Land Ores Versus Deep-Sea Polymetallic Nodules," *Journal of Cleaner Production* 275 (2020): 1–20; D. Paulikas et al., "Deep-Sea Nodules Versus Land Ores: A Comparative Systems Analysis of Mining and Processing Wastes for Battery-Metal Supply Chains," *Journal of Industrial Ecology*, https://doi.org/10.1111/jiec.13225.

27. A. N. Zerbini, G. Adams, J. Best, et al., "Assessing the Recovery of an Antarctic Predator from Historical Exploitation," *Royal Society Open Science* 6 (2019), http://doi.org/10.1098/rsos.190368.

28. A. Chin and K. Hari, "Predicting the Impacts of Mining of Deep Sea Polymetallic Nodules in the Pacific Ocean: A Review of Scientific Literature," *Deep Sea Mining Campaign and MiningWatch Canada* (2018): 1–60, http://www.deepseaminingoutofourdepth.org/wp-content/uploads/Nodule-Mining-in-the-Pacific-Ocean-1.pdf.

29. D. Kriebel, J. Tickner, P. Epstein, et al., "The Precautionary Principle in Environmental Science," *Environmental Health Perspectives* 109, no. 9 (2001): 871–76, https://www.ncbi.nlm.nih.gov/pmc/articles/PMC1240435/pdf/ehp0109-000871.pdf.

30. See, for example, D. Kennedy and V. Tauli-Corpuz, "Native Reluctance to Join Mining Industry Initiatives," *Culture Survival Quarterly*, April 2001.

31. H. Zehr et al., *The Big Book of Restorative Justice: Four Classic Justice and Peacebuilding Books in One Volume* (New York: Good Books, 2015).

EPILOGUE: GOVERNING OUR PLANET'S ELEMENTAL RESOURCES

1. T. Webb, "The War of the Four Oligarchs," *Observer*, April 13, 2008, https://www.theguardian.com/business/2008/apr/13/mining.russia1.

2. Lora Kolodny, "Tesla Has Bought Aluminum from Russian Company Rusal Since 2020, Showing How War Complicates Supply Chain," CNBC, March 14, 2022, https://www.cnbc.com/2022/03/14/tesla-has-bought-aluminum-from-russian-supplier-rusal-since-2020.html.

3. "Chinese Billionaire-Linked Aluminum Fraud Fined $1.8 Billion by U.S.," *Bloomberg.Com*, April 12, 2022, https://www.bloomberg.com/news/articles/2022-04-12/chinese-billionaire-linked-aluminum-fraud-gets-1-8-billion-fine.

4. Aluminum Stewardship Initiative, "Overview and History," https://aluminium-stewardship.org/about-asi/asi-history.

5. Aluminium Stewardship Initiative, "Issue Brief: Low Carbon Aluminium," https://aluminium-stewardship.org/low-carbon-aluminium.

6. Details of the theoretical and practical energy discussions are from U.S. Department of Energy, "U.S. Energy Requirements for Aluminum Production," https://www1.eere.energy.gov/manufacturing/resources/aluminum/pdfs/al_theoretical.pdf.

7. Vedanta ranks in the top 10 in the Dow Jones Sustainability Index: *Manufacturing India Today*, February 15, 2022, https://www.manufacturingtodayindia.com/sectors/vedanta-aluminum-makes-it-into-dow-jones-sustainability-index-2021-top-10-rankings.

8. "EGA and GE to Develop Roadmap to Decarbonise UAE Aluminum Giant's GE Gas Turbines, Including by Switching to Hydrogen," EGA, November 28, 2021, https://www.ega.ae/en/media-releases/2021/november/ega-and-ge-release/.

9. M. Bevir, ed., *Encyclopedia of Governance* (Thousand Oaks, CA: Sage, 2007).

10. "The Role of Critical Minerals in Clean Energy Transitions—Analysis," IEA, https://www.iea.org/reports/the-role-of-critical-minerals-in-clean-energy-transitions.

11. Ernst Ulrich von Weizsäcker and Anders Wijkman, *Come On! Capitalism, Short-Termism, Population, and the Destruction of the Planet* (Springer, 2017).

12. Systemadmin_Umwelt, "Factor X," *Umweltbundesamt*, June 26, 2012, https://www.umweltbundesamt.de/en/topics/waste-resources/resource-conservation-in-the-manufacturing/factor-x.

13. M. Kimball, "The Aluminum Rule," *Medium* (blog), December 24, 2015, https://medium.com/@mileskimball/the-aluminum-rule-8603cbd5331e.

INDEX

aluminum, xiii–xiv; abundance of, 18, 58, 62; in airplanes, 112, 118; Alzheimer's disease and, 49–50; in antiperspirants, 41, 49; in automotive sector, 135, 156–57; in batteries, 162–64; biotic forms lacking, 39–40; bonding of, xiv; cancer concerns with, 49; cartels for, 26; cells removing, 48; in chondrite, 15; concentration factor for, 92, *93*; confluence of transatlantic discovery and, 87–93, *93*; corrosion lacking with, 80–81; in cosmetics, 49; demand for, xvi; demographic power and upscaling innovation in, 230–34, *233*; in durability dilemma, 186–93, *190*; in Earth's crust, 15–18, 78; electrolytes for, 87; electron transfer of, 36; electron valence of, 40; elemental liberators and, 45–52; elixirs in, 39–45, *41*, *44*; emissions from, 184–85; environmental availability of, 53; in fish, 52–53; fortune of foundries for, 94–100; French interest in, 81; global cycle of, xv–xvi, *xvii*; glocalization of material governance and, 253–57; in green transition, 161–62; health activism around, 44–45; health impact of, 46–47; hydrogen from, 157–58; in hydro-metal nexus, 181–86; hydropower and, 181–84;

industrial synthesis of, 83–84, *84*; as inscrutable, 62; isolation of, 63–64; isotope tracing of, 50; in kitchen products, 150, 156; life emergence and, 29, 39; life rejecting, 35–36, 39–40, 48; as lithophile, 16; in material diaspora, 174–81; materiality of, 243–44; as metal, 78–80; metallic combinations in age of high technology and, 154–65, *159*; as metalloid, 17; in mineral form, 78; mineral nationalism, economic development and, 136–44; as miracle metal, 76–82; monopoly power, military-industrial complex and, 128–36; native, 77–78; oligopoly innovation and, 149–54; peacetime uses of, 123–24; on periodic table, 16–17, *17*; physical forms of, 18–19; in plants, 48; political economy influenced by, 135–36; as pollutant, 52; as precious metal, 84–85; primacy of, 14–19, *17*; production of, 82, 85, 87–92; prominence of, 81–82; purple purpose and, 19–23; in quasicrystals, 158–59, *159*; recycling of, 185, 200; in rocket fuel, 123–24; in satellites and spacecraft, 124–25; scaling process for, 79; scandium and, 160–61; silicon and, 52–58, 78; social restoration challenge for,

for, 26; chemical collusion and, 23–29; Henry VIII cut ties with, 28; Medici contract with, 25–27

Cavendish, Henry, 73–74

Chaloner, Thomas, 28

Chance, Stephen, 29

Chaney, Rufus L., 225–26

Channichthyidae (white-blooded ice fish), 70–71

Cherney, Michael, 246

Chernyshevsky, 81

Chester, Hamilton, 79

China, xv–xvi; aluminum stockpile from, 247–48; mineral sector in, 230; REE export restrictions in, 230–31; WTO on, 230–31

Chinese University of Geosciences, 231

chirality, 57

chondrite, 15

Churchill, Winston, 123

circular economy, 157; batteries in, 163–64; as complex system, 189, *190*; employment in, 193; human development impacted by, 192–93; Jevons' paradox in, 190–91, 194; material flow in, 185, 189; natural systems and, 192; principles of, 256; software in, 196; subsistence economies and, 193; systems thinking and, 191–92; tradeoffs in, 174. *See also* planned obsolescence; recycling

Clapp, George Hubbard, 91

Clark, Helen, 142

cobalt: as critical commodity, 227; phytomining of, 228

Codes of Justinian, 136–37

Cohen, Morris, 241

color: of elements, 19; in human apparel, 20; purple purpose and, 19–23

combustion: alchemy on, 71–72; chemistry of, 71; conundrum of, 71–76; of metals with oxides, 74–75; oxygen in, 71, 75; phlogiston theory on, 71–72; thermite reactions in, 75–76; weight changes after, 74

Committee on the Study of Materials (COSMAT), 241

complex adaptive system, 188–89, *190*

concentration factor, 92, *93*

Condé, Alpha, 141–42

Coors, William K., 173

corpuscules: aluminum atoms in, 78; matter composed of, 76–77

COSMAT. *See* Committee on the Study of Materials

COVID-19 pandemic, 155, 164

Cronstedt, Axel Fredrik, 165–66

cryolite: in aluminum production, 100, 105–7; arctic mine and synthetic sunset of, 100–108; composition of, 100; melting point of, 89; mining of, 103–7, *105*; naming of, 101; sodium carbonate from, 101, 103–4; synthetic, 105

CT. *See* carbothermic reduction

Culver, David, 135